Anomaly Detection - Recent Advances, AI and ML Perspectives and Applications

Edited by Venkata Krishna Parimala

Published in London, United Kingdom

Anomaly Detection - Recent Advances, AI and ML Perspectives and Applications
http://dx.doi.org/10.5772/intechopen.110988
Edited by Venkata Krishna Parimala

Contributors
Menachem Domb, Sujata Joshi, Arulmozhi Khn, Surendra Bhosale, Achala Deshmukh, Bhushan Deore, Parag Bhosale, Farrukh Arslan, Aqib Javaid, Muhammad Danish Zaheer Awan, Ebad-ur-Rehman, Siamak Parhizkari, Hironori Uchida, Yoshihisa Nakatoh, Yujle Li, Keitaro Tominaga, Hideki Itai, Miloš Cekić, Venkata Krishna Parimala

First published in London, United Kingdom, 2024 by IntechOpen
IntechOpen is the global imprint of INTECHOPEN LIMITED, registered in England and Wales, registration number: 11086078, 5 Princes Gate Court, London, SW7 2QJ, United Kingdom

British Library Cataloguing-in-Publication Data
A catalogue record for this book is available from the British Library

Additional hard and PDF copies can be obtained from orders@intechopen.com

Anomaly Detection - Recent Advances, AI and ML Perspectives and Applications
Edited by Venkata Krishna Parimala
p. cm.

This title is part of the Artificial Intelligence Book Series, Volume 24
Topic: Applied Intelligence
Series Editor: Andries Engelbrecht
Topic Editor: Carlos M. Travieso-Gonzalez

Print ISBN 978-1-83769-026-8
Online ISBN 978-1-83769-027-5
eBook (PDF) ISBN 978-1-83769-028-2
ISSN 2633-1403

For EU product safety concerns:
IN TECH d.o.o., Prolaz Marije Krucifikse Kozulić 3, 51000 Rijeka, Croatia,
info@intechopen.com or visit our website at intechopen.com.

IntechOpen Book Series
Artificial Intelligence
Volume 24

Aims and Scope of the Series

Artificial Intelligence (AI) is a rapidly developing multidisciplinary research area that aims to solve increasingly complex problems. In today's highly integrated world, AI promises to become a robust and powerful means for obtaining solutions to previously unsolvable problems. This Series is intended for researchers and students alike interested in this fascinating field and its many applications.

Meet the Series Editor

Andries Engelbrecht received the Masters and Ph.D. degrees in Computer Science from the University of Stellenbosch, South Africa, in 1994 and 1999 respectively. He is currently appointed as the Voigt Chair in Data Science in the Department of Industrial Engineering, with a joint appointment as Professor in the Computer Science Division, Stellenbosch University. Prior to his appointment at Stellenbosch University, he has been at the University of Pretoria, Department of Computer Science (1998-2018), where he was appointed as South Africa Research Chair in Artifical Intelligence (2007-2018), the head of the Department of Computer Science (2008-2017), and Director of the Institute for Big Data and Data Science (2017-2018). In addition to a number of research articles, he has written two books, Computational Intelligence: An Introduction and Fundamentals of Computational Swarm Intelligence.

Meet the Volume Editor

Dr. Venkata Krishna Parimala is currently a Professor of Computer Science and Director for University Rankings at Sri Padmavati Mahila University, Tirupati, India. He received his BTech in Electronics and Communication Engineering from Sri Venkateswara University, Tirupathi, India; an MTech in Computer Science and Engineering from Regional Engineering College (REC), Calicut, India; and a Ph.D. from VIT University, Vellore, India. Dr. Krishna has several years of experience working in academia, research, teaching, consultancy, academic administration, and project management roles. His current research interests include mobile and wireless systems, data science, explainable artificial intelligence (XAI), machine learning, and cloud computing. He is the recipient of several academic and research awards, such as the Cognizant Best Faculty Award for 2009–2010 and the 2021 Best Researcher Award from Sri Padmavati Mahila University. Recently Prof. Krishna was listed among the top 2% of scientists in the world by Stanford University, USA. He has authored more than 200 research papers in various national and international journals and conferences. He has delivered several keynote addresses and chaired sessions at reputed conferences. He is a senior member of several professional societies such as the Institute of Electrical and Electronics Engineers (IEEE), Association for Computing Machinery (ACM), Computer Society of India (CSI), Institution of Engineers (India) (IE(I)), and others.

Contents

Preface

Anomaly detection is the practice of pinpointing outliers in a set of similar items. Its main objective is to discern between typical and atypical data. Various strategies, including statistical techniques, supervised and unsupervised learning, semi-supervised learning, time-series analysis, and deep learning (DL) methodologies, can be employed to perform anomaly detection. Anomaly detection is versatile, finding use in diverse domains such as medicine, finance, manufacturing, identity protection, data security, networking, video surveillance, and cyber security.

The emergence of next-generation networks has brought forth an array of sophisticated and complex systems that are often vast, decentralized, and dynamic. These networks, which might include 6G and beyond, Internet of Things (IoT) networks, or even intricate data center networks, have multifaceted traffic patterns, a multitude of interconnected devices, and varied user behaviors. Such complexity makes them susceptible to a wide range of anomalies, from performance hiccups to security breaches. Traditional anomaly detection methods might struggle with the sheer scale and intricacy of these networks. In contrast, machine learning (ML), with its capacity to learn from vast amounts of data, can potentially discern patterns and anomalies that would be imperceptible to conventional methods. Given the critical nature of networks, especially in sectors like healthcare or finance, it is paramount not just to detect anomalies but also to understand the "why" behind them. By making artificial intelligence's (AI) decisions interpretable, it becomes easier for network administrators or security experts to take informed actions, rectify issues, or even preemptively address potential vulnerabilities.

In light of the aforementioned context, the editor requested chapters that address various aspects of anomaly detection, such as novel strategies based on AI, machine learning, optimization, control, statistics, and social computing, among others. We received several chapter proposals, and each chapter was reviewed thoroughly. Three sections and seven chapters were finally selected for this book and the details of the chapters are as follows:

Chapter 1 by Venkata Krishna Parimala offers a foundational introduction to the book, establishing a solid framework for understanding anomaly detection from AI and ML perspectives. This chapter lays the groundwork by elucidating key concepts, methodologies, and the significance of anomaly detection in the realm of AI) and ML, setting the stage for more detailed explorations in subsequent chapters. The first section of the book deals with anomaly detection with time series of data. Chapter 2 presents a review of anomaly detection in medical time series with generative adversarial networks. The author, Miloš Cekić, discusses anomaly detection in medical time series, such as diagnosing diseases like epilepsy or preventing fatal events like cardiac arrhythmias. Generative adversarial networks (GANs) have demonstrated potential in various areas, including cybersecurity and data augmentation. Recently, they have been applied to detect anomalies in medical time series. This chapter reviews the use of GANs in this context, addressing the nature of time-series anomalies, challenges in medical time series, and DL issues. The discussion includes popular GAN models and their application in detecting anomalies in ECG and EEG medical time series. Chapter 3 discusses anomaly detection with time series in IoT. This chapter explores anomaly detection in IoT using ML and DL. With 85 billion devices anticipated by 2025, cyber security challenges arise, according to the authors Menachem Domb et al. The chapter reviews these issues, suggesting protocols and solutions for a safer IoT landscape. Chapter 4 explores the intricacies of anomaly detection in time series. The authors, Farrukh Arslan et al., delve into the numerous challenges anomaly detection confronts in contemporary applications.

The second section of the book discusses anomaly detection vs. intrusion detection. In Chapter 5, Surendra Bhosale et al. explore anomaly detection using the Adaptive Dolphin Atom Search Optimization (DASO) method. They utilize DASO combined with deep RNN techniques to address anomaly detection and intrusions. In Chapter 6, Siamak Parhizkari explains anomaly detection within intrusion detection systems. The chapter examines various facets of anomaly detection, including signature-based detection and both supervised and unsupervised learning methods. It further details the application of anomaly detection in intrusion detection systems.

The third section of the book presents anomaly detection models. In Chapter 7, Hironori Uchida et al. explain software log anomaly detection models. The chapter presents technological advancements in automated software log analysis. Despite DL's high accuracy in software log anomaly detection, its adoption in software development remains limited. Evaluations of five models, including the proposed Neocortical Algorithm, on the BGL dataset revealed overfitting tendencies and highlighted the need for diverse datasets.

This book will assist researchers in understanding the advancements occurring in the field of anomaly detection with AI and ML techniques and their applications.

I would like to thank Publishing Process Manager Ms. Karla Skuliber and other members of the editorial team at IntechOpen for their kind cooperation and help. I also extend our sincere thanks to the contributing authors and reviewers for their interest and support.

Venkata Krishna Parimala
Computer Science Department,
Sri Padmavati Mahila University,
Tirupati, India

Section 1

Anomaly Detection with Time Series Data

Chapter 1

Introductory Chapter: Anomaly Detection – Recent Advances, AI and ML Perspectives and Applications

Venkata Krishna Parimala

1. Introduction

The significance of anomaly detection transcends industries and impacts various facets of daily life and societal functioning. In the world of finance, it serves as a guardian of economic stability. Beyond fraud detection, it helps regulatory authorities monitor for signs of market manipulation or systemic risks that could lead to economic downturns. It is not just about protecting individual investors; it's about safeguarding the entire financial infrastructure on which modern economies rely.

In healthcare, the stakes are even more personal. Anomaly detection algorithms are being integrated into wearable devices, constantly monitoring physiological data to provide real-time health insights. This has the potential to revolutionize preventive medicine by catching symptoms before they manifest into more severe conditions, thereby facilitating early intervention and potentially saving lives.

In transportation, particularly in aviation and autonomous vehicles, anomaly detection is critical for ensuring safety. Algorithms continuously monitor system health and can alert human operators or initiate fail-safes if something goes awry. The ability to detect a malfunction before it leads to a catastrophic failure could mean the difference between a controlled emergency landing and a tragic accident.

The technology also has growing applications in environmental protection. Algorithms can analyze satellite imagery to identify illegal deforestation or poaching activities, enabling timely intervention. Similarly, in marine biology, anomaly detection helps researchers identify unusual patterns in sea temperature or marine life behavior, offering early indicators of environmental issues like ocean acidification.

Additionally, anomaly detection plays a critical role in the realm of data integrity and information verification. In the age of 'fake news,' these algorithms can sift through vast amounts of data to flag misinformation or anomalous reporting, thereby helping to maintain the integrity of public discourse.

Finally, the technology is making inroads into the field of disaster management. By analyzing data from seismic sensors, weather satellites, and historical records, anomaly detection can provide early warnings for natural disasters like earthquakes, tsunamis, or hurricanes, enabling timely evacuations and preparation, thereby minimizing loss of life and property.

IntechOpen

The significance of anomaly detection is multi-dimensional, affecting both individual lives and the larger fabric of society. Its potential to drive proactive solutions, prevent crises, and even save lives makes it an indispensable tool in the modern data-driven world.

2. The limitations of traditional methods

Traditional methods of anomaly detection have provided a foundational framework for identifying outliers in data, but as data have grown more complex, these methods are showing their limitations more prominently. One of the most glaring issues is the assumption of a specific data distribution. Traditional techniques often assume that data follow a Gaussian or similar distribution, an assumption that is frequently violated in real-world applications. This not only affects the accuracy but also limits the type of anomalies that can be detected.

Another substantial limitation is scalability. Traditional methods were not designed to handle the massive datasets generated in contemporary applications, such as social media analytics, sensor networks, and large-scale e-commerce. Processing large datasets often requires significant computational resources, making these methods inefficient and sometimes impractical for big data applications.

Sensitivity to parameter settings is another drawback. The effectiveness of traditional methods often hinges on the appropriate selection of parameters like thresholds or cluster sizes. Inconsistent or suboptimal parameter selection can result in missed anomalies or an excessive number of false alarms. This makes traditional methods highly dependent on domain expertise and often requires manual tuning, which is both time-consuming and susceptible to human error.

Traditional methods also struggle with high-dimensional data. In scenarios where multiple attributes or features are involved, the effectiveness of traditional methods diminishes. They often suffer from the "curse of dimensionality," a phenomenon where the data become increasingly sparse as the dimensionality increases, making it challenging to identify meaningful patterns.

The issue of temporal dynamics is another limitation. Traditional methods are often ill-suited for detecting anomalies in time-series data where temporal correlations are essential. They usually treat data points as independent entities, ignoring the temporal relationships that are often crucial for accurate anomaly detection in sequences.

Lastly, interpretability and transparency, although considered a strength of traditional methods, can also be a limitation. The simplified models may offer easier interpretation but at the cost of capturing the complexities of the data. This trade-off often leads to models that are overly simplistic, failing to capture the nuanced behaviors that more advanced models can identify.

3. The role of AI and ML in anomaly detection

The infusion of artificial intelligence (AI) and machine learning (ML) technologies into anomaly detection is revolutionizing the field, offering a robust set of tools and methodologies that far exceed the capabilities of traditional techniques. These advanced algorithms are designed to tackle multi-dimensional and large-scale data, making them well-suited for modern applications that often involve big data and streaming analytics.

Machine learning models like Random Forests and Support Vector Machines have been particularly effective in feature selection and reducing dimensionality, which are common challenges in high-dimensional data spaces. Deep learning techniques, such as Long Short-Term Memory (LSTM) networks, have shown exceptional performance in time-series anomaly detection, a critical aspect in sectors like finance and industrial automation. More recently, Generative Adversarial Networks (GANs) have been adapted for anomaly detection, proving effective in learning complex data distributions without the need for explicit labeling.

One of the most compelling advancements is the introduction of semi-supervised and unsupervised learning techniques. These models do not require a fully labeled dataset for training, a feature that is particularly advantageous in scenarios where labeling is costly or impractical. This opens up new avenues for anomaly detection in fields like cybersecurity, where attacks are continually evolving, and manual labeling quickly becomes obsolete.

Furthermore, the AI and ML models are increasingly becoming capable of real-time learning, a critical requirement in dynamic environments. For example, reinforcement learning algorithms can interact with their environment in real-time, adapting their anomaly detection strategies as they gain more information. This is invaluable in applications such as autonomous driving and real-time network security, where the cost of failing to detect an anomaly could be catastrophic.

In addition to performance benefits, AI and ML are also contributing to the explainability and interpretability of anomaly detection models. With the advent of techniques like Local Interpretable Model-agnostic Explanations (LIME) and SHAP (SHapley Additive exPlanations), these complex models are becoming less of a 'black box,' thereby gaining greater acceptance in fields that require rigorous validation, such as healthcare and aviation.

Anomaly detection is a growing field with applications across various domains such as healthcare, building management, cybersecurity, weather forecasting, and surveillance. With the advent of artificial intelligence (AI) and machine learning (ML), sophisticated techniques are being developed to tackle complex anomaly detection tasks. However, each domain has its own set of challenges and requirements that influence the choice of techniques and their effectiveness.

In healthcare, Cekić et al. [1] shed light on the importance of anomaly detection in medical time series data, such as electrocardiography (ECG) and electroencephalography (EEG). They highlight the use of Generative Adversarial Networks (GANs) for this purpose. While GANs have shown promise, they also present challenges related to medical data, such as limited labeled samples and the complex nature of anomalies. In a similar vein, Esmaeili et al. [2] investigate the use of GANs for anomaly detection in biomedical imaging. Their study, conducted on seven different medical imaging datasets, shows highly variable performance (AUC: 0.475-0.991; Sensitivity: 0.17-0.98; Specificity: 0.14-0.97), indicating the method's limitations and the need for further research.

In the context of building management, Copiaco et al. [3] take a unique approach by using two-dimensional (2D) image representations of energy time-series data for deep anomaly detection. Their method achieved impressive F1-scores of 93.63 and 99.89% on simulated and real-world datasets, respectively. Himeur et al. [4] expand on this by surveying AI and big data analytics in building automation and management systems (BAMSs). They identify the current limitations, including the system's focus primarily on heating, ventilation, and air conditioning (HVAC) controls, and suggest AI as a promising solution.

Cybersecurity is another critical application area. Javaheri et al. [5] focus on Distributed Denial of Service (DDoS) attacks, providing a comprehensive survey that proposes effective defensive strategies. They emphasize the use of fuzzy logic-based methods as a promising avenue for future research. Zehra et al. [6] discuss the security challenges in Network Function Virtualization (NFV), advocating for machine learning-based anomaly detection techniques to enhance network security.

In other specialized applications, Jin et al. [7] provide a comprehensive review of Graph Neural Networks (GNNs) for time series analysis, which includes forecasting, classification, and anomaly detection. Their work serves as a guide to understand the strengths and limitations of using GNNs for time-series data. Patriarca et al. [8] delve into the importance of weather forecasting for aerodrome operations and propose a machine learning-based approach for anomaly detection in historical weather data. Finally, Şengönül et al. [9] explore the use of AI in surveillance video anomaly detection, noting the increasing need for automated systems due to the sheer volume of video data being generated.

In summary, while AI and machine learning offer promising solutions for anomaly detection across domains, the effectiveness of these techniques varies significantly. The limitations often arise from domain-specific challenges such as data sparsity, complexity of the anomalies, and computational constraints. Therefore, tailored approaches and continuous research are essential for advancing the field.

Author details

Venkata Krishna Parimala
Computer Science Department, Sri Padmavati Mahila University, Tirupati, India

*Address all correspondence to: pvk@spmvv.ac.in

References

[1] Cekić M. Anomaly Detection in Medical Time Series with Generative Adversarial Networks: A Selective Review. London: IntechOpen; 2023. DOI: 10.5772/intechopen.112582

[2] Esmaeili M et al. Generative adversarial networks for anomaly detection in biomedical imaging: A study on seven medical image datasets. IEEE Access. 2023;**11**:17906-17921. DOI: 10.1109/ACCESS.2023.3244741

[3] Copiaco A, Himeur Y, Amira A, Mansoor W, Fadli F, Atalla S, et al. An innovative deep anomaly detection of building energy consumption using energy time-series images. Engineering Applications of Artificial Intelligence. 2023;**119**:105775. DOI: 10.1016/j.engappai.2022.105775

[4] Himeur Y, Elnour M, Fadli F, et al. AI-big data analytics for building automation and management systems: A survey, actual challenges and future perspectives. Artificial Intelligence Review. 2023;**56**:4929-5021. DOI: 10.1007/s10462-022-10286-2

[5] Javaheri D, Gorgin S, Lee J-A, Masdari M. Fuzzy logic-based DDoS attacks and network traffic anomaly detection methods: Classification, overview, and future perspectives. Information Sciences. 2023;**626**:315-338. DOI: 10.1016/j.ins.2023.01.067

[6] Zehra S, Faseeha U, Syed HJ, Samad F, Ibrahim AO, Abulfaraj AW, et al. Machine learning-based anomaly detection in NFV: A comprehensive survey. Sensors. 2023;**23**:5340. DOI: 10.3390/s23115340

[7] Jin M, Koh HY, Wen Q, Zambon D, Alippi C, Webb GI, et al. A survey on graph neural networks for time series: Forecasting, classification, imputation, and anomaly detection. 2023. arXiv:2307.03759 [cs.LG]. DOI: 10.48550/arXiv.2307.03759

[8] Patriarca R, Simone F, Di Gravio G. Supporting weather forecasting performance management at aerodromes through anomaly detection and hierarchical clustering. Expert Systems with Applications. 2023;**213**(Part C):119210. DOI: 10.1016/j.eswa.2022.119210

[9] Şengönül E, Samet R, Abu Al-Haija Q, Alqahtani A, Alturki B, Alsulami AA. An analysis of artificial intelligence techniques in surveillance video anomaly detection: A comprehensive survey. Applied Sciences. 2023;**13**(8):4956. DOI: 10.3390/app13084956

Chapter 2

Anomaly Detection in Medical Time Series with Generative Adversarial Networks: A Selective Review

Miloš Cekić

Abstract

Anomaly detection in medical data is often of critical importance, from diagnosing and potentially localizing disease processes such as epilepsy to detecting and preventing fatal events such as cardiac arrhythmias. Generative adversarial networks (GANs) have since their inception shown promise in various applications and have been shown to be effective in cybersecurity, data denoising, and data augmentation, and have more recently found a potentially important place in the detection of anomalies in medical time series. This chapter provides a selective review of this novel use of GANs, in the process highlighting the nature of anomalies in time series, special challenges related to medical time series, and some general issues in approaching time series anomaly detection with deep learning. We cover the most frequently applied GAN models and briefly detail the current landscape of applying GANs to anomaly detection in two commonly used medical time series, electrocardiography (ECG) and electroencephalography (EEG).

Keywords: anomaly detection, medical time series, generative adversarial network (GAN), electrocardiogram (ECG), electroencephalogram (EEG)

1. Introduction

The increasingly widespread deployment of advanced technology in modern healthcare has led to exponential growth in the generation and collection of medical time series data, which offer unprecedented opportunities for the detection and diagnosis of disease processes. Being able to use these "big" data to detect and localize anomalies can lead to early and precise disease recognition, timely and proactive intervention, development of personalized treatment plans, improved patient outcomes, and better risk management while at the same time offering opportunities for better understanding and classification of disease pathology [1–5].

The extraction of meaningful insights from medical time series data, however, poses a number of significant challenges. For one, the data are highly complex and often multidimensional/multimodal and nonstationary. They also usually suffer from noise, missing values, and artifacts as well as from the inherent variability of human physiology and broad range of normality across individuals, which makes the identification of "abnormal" a nontrivial task [6]. Anomalies within the data can be highly heterogeneous and can manifest as subtle deviations or sudden, drastic changes. Additionally, specific disease data that could be used to train analytical models are at best highly heterogeneous and more likely simply not available; even when data are available, properly labeled data are generally lacking. Finally, at the present time, there is no expert knowledge related to large medical datasets that can parallel classical medical knowledge—it is sometimes not even clear what exactly to look for since traditional medical semiology operates in a different conceptual space (defined signs/symptoms vs. patterns in massive data) [1–3, 6].

Traditionally, statistical and rule-based methods have been employed for anomaly detection in medical time series data [4, 7, 8]. These methods rely on predefined thresholds, statistical models, or expert knowledge to identify deviations from normal. These approaches, however, often struggle to capture complex and nonlinear patterns that may be present in the data. This has recently led to a growing interest in leveraging machine learning and deep learning techniques for anomaly detection in medical imaging [1–3, 6, 7, 9–11]. A widely used type of model is the generative adversarial network (GAN), which has demonstrated superiority in a variety of tasks in medical imaging due to its powerful ability to learn the distribution of the training data and to generate novel but realistic samples that reflect the underlying data characteristics [12–26]. A GAN trained on normal instances only, for example, can capture the complex patterns and dependencies inherent in the data, enabling the generation of synthetic samples that closely adhere to the learned distribution. Anomalies, being significantly different from normal, can then be identified by how far they deviate from their reconstruction [27–32]. The application of GANs in anomaly detection for medical time series has demonstrated promising results, and GAN-based approaches have been shown to be able to effectively capture temporal dependencies, handle complex patterns, and adapt to individual patient variations. Moreover, they can detect subtle anomalies that may go unnoticed by traditional methods [29, 31, 32].

This chapter endeavors to provide a brief review of the current landscape of the use of GANs in anomaly detection in medical time series data. We first present a brief overview of properties of anomalies, time series data, and specific challenges related to medical time series. We then discuss general ways of approaching time series with deep learning methods before discussing GANs and GAN applications to anomaly detection in general and to time series in particular. We then review the current state of the use of GANs in medical imaging and anomaly detection in specific fields of electrocardiography (ECG) and electroencephalography (EEG). Finally, we briefly discuss some challenges and future directions.

2. Anomalies in time series: problem complexities and challenges

2.1 Properties of time series data

While time is a fundamental concept in nearly all data, time series explicitly involve the temporal dimension. The following is a brief summary of the specific properties of time series and how they affect anomaly detection.

2.1.1 Temporality

A time series is an ordered sequence of data points indexed by time (usually but not always across equal temporal intervals) [33–36]. We can define a time series as a vector X such that: $X = \{x_1, x_2, \dots, x_t\}$, where x_i represents the datum at time $i \in T$ and $T = \{1, 2, \dots, t\}$. The (necessary) assumption of continuity of the underlying generative process implies that each point is in some way conditioned on previous values (past states of the process), with this dependency captured as a joint distribution of a set of observations: $p(x_1, x_2, \dots, x_t) = p(x_1) \prod_2^t p(x_t | x_1, x_2, \dots, x_{t-1})$. The influence of the past is generally assumed to decrease with time, though this may not necessarily be the case [33].

2.1.2 Dimensionality

Time series data may be univariate or multivariate, with the dimension representing the number of individual data attributes captured at each time point. The above specification of a time series vector is univariate. Multivariate series can be defined as a time-ordered set of multidimensional vectors X_t (rather than points), with $X_t = \left(x_t^1, x_t^2, \dots, x_t^d\right)$, where d is the number of dimensions; the multivariate time series is then a rank d + 1 tensor X_i^j, where $j \in D$ and $D = \{1, 2, \dots, d\}$ is the number of dimensions and $i \in T$ and $T = \{1, 2, \dots, t\}$ denotes time [9, 33]. Alternatively, multivariate series can also be conceptualized as a collection of univariate time series. While analysis of univariate time series needs to consider only the relationship between the current state and previous states (temporal dependency), multivariate series entail dependencies and correlations (semantics) across both previous states (temporal) within a series and other dimensions (spatial) at any given time point, keeping in mind that any given datum may also depend on a mixture within and across different time series (spatiotemporal dependencies). These dependencies may be multiscale (short-, medium-, or long-range) and in some cases nonstationary or dynamic, meaning that the scale and structure of dependencies itself may vary in time [37–40].

2.1.3 Nonstationarity

A time series is assumed to be stationary if its statistical properties do not change over time. Most real-world time series are not stationary, however, meaning the mean and variance (and other moments of the distribution) vary. Common sources of nonstationarity include trends (baseline drift that may be local or global and linear or nonlinear), seasonal cycles (with a stable period), nonseasonal cycles (with a variable period), pulses and steps (including concept drift and change points, instances where the relationship between input and output changes), and random/irregular movement. Because nonstationarity implies that the data distribution itself changes in some way, an appropriate model will need to somehow capture the underlying generative process rather than the statistics of the apparent data [33–36].

2.1.4 Noise

Real-world datasets typically contain a significant amount of noise or unwanted signal, which represents the semantic boundary between normal data and true anomalies. The classical definition of an anomaly or outlier is "an observation which deviates so much from other observations as to arouse suspicions that it was generated by a different mechanism" [41]. In this context, it is helpful to differentiate between outlier and anomaly: *outlier* refers to "unusual data objects that are statistically rare but may or may not result from the same generative process as other data objects," while, in contrast, *anomalies* are defined as "data objects that have resulted from a distinct generative process that may or may not appear in the data as outliers" [42]. This distinction is necessarily contextual and application-specific, although typically anomalies will have a much higher outlier score than noise [43] since they are presumably generated by a different underlying process.

2.2 Properties of anomalies

The detection of anomalies is a specific problem in pattern recognition that is distinct from other analytical and learning tasks. Key complexities are discussed below. These issues apply to any anomaly detection and not just time series, but with time series and medical time series in particular additional specific challenges arise.

2.2.1 Unknown nature of anomalies

Anomalies are by definition unknown and may involve unknown abrupt behaviors, patterns, or distributions, which remain unknown until they occur. Even if a particular type of anomaly is known and categorized (there are two distinct issues here, since even if an anomaly type exists, it may not be immediately classifiable due to the heterogeneity of its manifestations), recognizing it may still be difficult. With machine learning, there is the added complexity that, even if various diseases are categorized in terms of their specific symptoms, the disease process may not be clearly defined within the particular modality performed for its detection [10, 11, 33, 37–39].

2.2.2 Anomaly class heterogeneity

Since anomalies are irregular and heterogeneous, one class of anomalies may have very different abnormal characteristics when compared to another class of anomalies; in other words, not only are anomaly classes themselves heterogeneous, the heterogeneity *within* anomaly classes or types is itself heterogeneous [11, 36, 39].

2.2.3 Dataset/class imbalance

Anomalies are typically rare events, occurring much less often than normal instances, which account for the overwhelming majority of the data. It is therefore extremely difficult and labor-intensive (if not impossible) to collect a sufficient amount of labeled and/or clearly defined abnormal instances that could be used for anomaly definition and model training. The result is severe class imbalance in any potential training set [29, 32, 40].

2.2.4 Types of anomalies

Anomalies can generally be classified into three types, point anomalies, contextual anomalies, and collective anomalies, which in time series correspond to abnormal time points, abnormal time intervals/subsequences, and abnormal time series [10, 29, 33, 37, 39].

2.2.4.1 Point anomalies

A point anomaly is a single datum at which the value of the observed variable is significantly different either from the entire time series (global) or from neighboring points in a time series (local or contextual). Point anomalies may be univariate or multivariate and usually entail extreme values. An example may include an abnormal blood pressure reading, which would then need to be defined as noise or something potentially indicating a deeper problem.

2.2.4.2 Time interval/subsequence anomalies

This type involves a subsequence of points that does not reflect the normal behavior of the system and in which each individual observation may be within normal range but the subsequence as a whole is anomalous. The subsequence may affect a single (univariate) or multiple (multivariate) time-dependent variables. An example would be an epileptic seizure on electroencephalogram (EEG), which may not be out of normal range in any individual point, but in which the multivariate pattern across multiple electrodes over a finite time period reflects the abnormality.

2.2.4.3 Collective or time series anomalies

This class includes cases where the entirety of a (or several) time series of a multivariate dataset is anomalous with respect to the dataset as a whole. This type is distinct from subsequence anomalies due to the length of the anomaly, which extends to the same length as the sequence. The idea here is that what is represented is not a temporary anomaly in the functioning of a part of a multidimensional system that returns to normal at some point, but rather that there is a persistent underlying abnormality in some portion of the system that can only be detected in the context of the entire dataset. An example would be abnormal signal related to a voxel (volume element) or group of voxels in a resting-state functional magnetic resonance imaging (MRI) (rs-fMRI) study representing a brain region that might reflect an underlying disease process such as an epileptogenic focus or tumor. The only way to detect this is by looking at the relationship of the various time series to each other both within individual patients and across individuals.

2.3 Challenges specific to medical time series

Anomaly detection in medical time series data comes with unique challenges, stemming from the nature of the data, the inherent variability in human physiology, and the requirement for the results to be interpretable by healthcare professionals. The following are the major domain-specific challenges.

2.3.1 Noise and artifacts

Medical time series data can contain a significant amount of noise due to sensor inaccuracies, patient movement, measurement errors, and physiological artifacts [1, 2]. For example, an ECG signal may contain noise from muscle contractions, or a glucose monitor may have inaccuracies due to calibration errors. These artifacts can distort the underlying physiological signal and lead to false detections of anomalies.

2.3.2 Missing and irregularly sampled data

Medical time series data often suffer from missing values and irregular sampling intervals. For example, a patient might remove a wearable device for a period, leading to missing data, or a sensor might malfunction. Irregularly sampled data can arise in outpatient settings where measurements are taken at each visit, but the visits occur at irregular intervals. These irregularities pose challenges for conventional time series analysis methods, which typically assume regular sampling, and require specialized techniques to handle missing values, synchronize timestamps, and ensure consistent analysis across different time series [1, 2, 6].

2.3.3 Nonstationarity

Medical time series data often exhibit nonstationarity, meaning that their statistical properties change over time. This could be due to a patient's changing health status, the effect of medications or interventions, or changes in external conditions such as time of day or physical activity. Traditional time series analysis techniques often assume stationarity, so nonstationarity poses a significant challenge [35, 43].

2.3.4 High dimensionality

Medical time series data can involve a high number of variables or channels, possibly with different types of data that are collected differently (e.g., blood pressure, heart rate, electrocardiogram, oxygen (O_2) saturation, and respiration may all be monitored simultaneously in the intensive care unit (ICU)) or may consist of massively multidimensional imaging data (e.g., for the open source dataset in the Human Connectome Project, each raw rs-fMRI time point contains 673,920 voxels or dimensions, which over the span of an approximately 15 min scan generates 8×10^8 data points per run per subject). This high dimensionality presents challenges in data storage, computation, visualization, and analysis. It also increases the need for complexity of anomaly detection algorithms, as they need to be able to handle and interpret data across multiple dimensions [1, 4, 5, 38].

2.3.5 Intra- and interpatient variability

There is a high degree of variability in physiological parameters both within the same individual over time (intrapatient variability) and between different individuals (interpatient variability). This variability complicates the task of defining what constitutes an "anomaly." An anomalous value for one individual might be normal for another, and a reading that is normal for a person at rest might be abnormal during exercise. Anomaly detection methods need to account for these interpatient variabilities and develop patient-specific or subgroup-specific models to accurately capture normal and abnormal patterns [1].

2.3.6 Lack of labeled data

Supervised learning methods, which can be very effective for anomaly detection, require labeled data for training. Obtaining such labeled data in healthcare settings is challenging, however, as it typically requires expert clinicians to manually review and label the data, which is time-consuming and expensive [2]. This essentially forces most anomaly detection (AD) models to be semi-supervised or unsupervised (see below).

2.3.7 Privacy and security

Medical data are highly sensitive, and strong safeguards are required to protect patient privacy. This can make it challenging to gather sufficient data for building robust anomaly detection models, and it also requires careful consideration when deploying these models to ensure that patient data are handled securely [1, 27, 28].

2.3.8 Interpretability and explainability

Healthcare professionals need to understand the detected anomalies and trust the reasoning behind them to make informed decisions. Anomaly detection methods should therefore provide interpretable results, visualizations, or explanations that can lead to improved or optimized treatment decisions. Many advanced machine learning models, however, and particularly deep learning models, often act as "black boxes," making their predictions difficult to interpret. This lack of interpretability can hinder the adoption of these models in clinical practice [1, 28].

2.4 Summary

Anomaly detection in multidimensional time series data is a significant challenge both from conceptual and technical points of view. Anomalies are by nature rare and heterogeneous, and therefore not easily amenable to classification, which means that they must generally be classified not as what they are but rather as what they are not. This makes not just detection but even simply definition of anomalies challenging: what is normal? How "different" must something be from some generalization (e.g., mean or average) of normal to constitute an anomaly? How do we define this difference? If the normal data change over time, how do we learn and describe this aspect of normality? What is the normal evolution of the behavior of our system? For multivariate time series, the issues multiply due to dependencies in the data that may extend over time and across multiple variables and may involve different spatiotemporal scales simultaneously. Determining normal behavior then is quite challenging, and the detection of potentially multiscale spatiotemporal anomalies even more so. As an extended example, consider rs-fMRI once again: say a brain region comprising a set of voxels is abnormal and therefore generates an anomalous signal. Detecting this anomaly requires that we are somehow able to learn all the multivariate dependencies at multiple scales across multiple subjects (the normal population): spatially between not just voxels but groups of voxels comprising brain regions, which may vary in size; temporally in that the length of the anomalous series may vary from short unusual behaviors to longer term abnormalities; and spatiotemporally in that different length signals from regions of different sizes may affect other regions at different spatial scales. Since anomalies are heterogeneous and we cannot know *a priori* exactly what we are looking for, detecting an "anomalous brain region" in the highly multidimensional rs-fMRI signal requires significant model complexity and computational power.

Finally, there are additional practical concerns: generalized unavailability of labeled training data, unreliable or noisy data, heterogeneously organized or collected data, data privacy (extremely important in medical datasets), interpretability, throughput and automatization, and scalability from specific research purposes to clinical applicability.

3. Anomaly detection in time series: concepts and models

3.1 Basic paradigms

Traditional anomaly detection methods in healthcare include statistical approaches, rule-based methods, and machine learning techniques. Statistical methods often rely on distribution-based models, such as Gaussian or Hidden Markov Models, to capture the normal patterns and detect deviations from them. More sophisticated classical models include autoregressive integrated moving average (ARIMA) models, exponential smoothing state space model (ETS), and Seasonal-Trend decomposition using LOESS (STL). These approaches are useful for detecting point or global anomalies, but are not helpful in the detection of contextual or collective anomalies [42–45] and are not useful for multivariate or multidimensional data [37–39]. Rule-based methods utilize expert-defined rules or thresholds to identify anomalies based on predefined criteria. Machine learning models, especially unsupervised machine learning algorithms, like clustering (K-means, density-based spatial clustering of applications with noise (DBSCAN)) or K-nearest neighbors (K-NN), are often used to detect outliers in time series data. Supervised learning methods, like support vector machine (SVM) or Random Forests, can also be used when labeled data are available. Deep learning methods, however, are best for capturing complex nonlinear relationships and dependencies in time series data [33, 35, 44–47]. Long Short-Term Memory (LSTM) networks and autoencoders (AEs) have been common choices due to their ability to model temporal dependencies, with GANs being utilized more recently for anomaly detection in time series [29, 33, 39].

Generally speaking, machine (and deep) learning methods can be grossly categorized along two axes: learning scheme vs. anomaly determination. Note that additional axes are possible, e.g., univariate vs. multivariate, anomaly type (point, subsequence, or time series); however, these will not be specifically considered here since most medical time series are multivariate and most deep learning approaches attempt to find anomalies in an anomaly-type agnostic manner [33].

3.1.1 Learning scheme

The three major learning schemes in machine learning are supervised, unsupervised, and semi-supervised. Supervised models aim to learn a mapping from data to their corresponding annotations and then use this mapping to perform classification on test data. Important examples of a supervised approach in the medical field are applications to automated brain tumor and ischemic stroke recognition and segmentation (using the extensively labeled Multimodal Brain Tumor Image

Segmentation, BraTS [48], and Ischemic Stroke Lesion Segmentation, ISLES, datasets [49]). Although these networks can successfully learn to recognize visual anomalies in medical images [2, 4], they require accurate labels and annotations and can only learn to recognize specific predefined abnormalities that are already present in the training set. Given the difficulty and time-consuming nature of labeling data manually, this approach is usually only appropriate for very specific use-cases.

Most approaches to anomaly detection time series therefore take the form of unsupervised or semi-supervised learning [32, 38, 45]. In unsupervised anomaly detection, the algorithms separate anomalies without prior knowledge or any explicit distinction between normal and abnormal, whereas in semi-supervised approaches all training data are assumed to be normal. Unsupervised methods are the most flexible since they depend entirely on the internal features of the data; however, this type of approach generates its own set of problems including potentially nonconvergent training, unclear recognition of anomalies, and difficulty in interpretation. For most medical data, including time series, a semi-supervised approach is generally considered to be most appropriate [6, 10, 11, 45–47], given the fact that the overwhelming amount of data collected is normal and that the anomalies themselves are highly heterogeneous.

3.1.2 Anomaly determination

Most anomaly detection involves an "anomaly score"—a number that is calculated based on the analytical technique that can then be compared to what is expected from a normal dataset. Once the data are learned (the model or distribution is assumed and fitted), a measure of the difference of each particular datum (whatever form this takes, point, subsequence, etc.) from the learned distribution must be determined, and this is the anomaly score. The following approach is nearly universal in machine (and by extension deep) learning for anomaly detection in time series: pick or create a model/architecture that we think will appropriately model the dataset, train the model on normal data in order to learn the data distribution (presumably with all dependencies), and then test new data with reference to the learned distribution using an anomaly score, which then determines if the datum in question was generated by the same underlying process the model was trained on or something different [11, 29].

Three basic approaches to determining anomaly scores are used, regardless of the time-modeling approach taken (see below): forecasting/prediction, reconstruction, and distance/dissimilarity [45].

3.1.2.1 Prediction

Prediction or forecasting-based models learn to predict expected future values based on the learned data, with anomalies determined based on the residual between the predicted value and the observed quantity. Most forecasting models use a sliding window to forecast one point at a time, although short sequences can also be generated. There is no robust forecasting-based model for rapidly and continuously changing time series, however, since such time series can only be predicted in the very short term if at all, and forecasting models are known to demonstrate significantly increased prediction errors as the number of time points increases [39]. This also makes them generally unsuitable for subsequence anomalies. Even in the deep learning context, forecasting-based models can only make short-term predictions with acceptable accuracy.

There is certainly a place for forecasting-based models in medical time series analysis, however. For any real-time applications where early or real-time anomaly detection is important, forecasting models are crucial. An example would be patients on telemetry or in the ICU, where blood pressure, respiration, and ECG may, individually or in combination, signal an impending collapse. This may also be the case in predicting the onset of seizures with EEG or cardiac arrest with ECG, where any advance notice of a critical event may significantly alter outcomes.

3.1.2.2 Reconstruction

Reconstruction-based models are not subject to the constraints of prediction models. With this type of model, normal behavior is modeled by encoding subsequences of normal training data (usually input as a sliding window that provides the temporal context for each datum) into a lower-dimensional latent space. In a semi-supervised context where the model is trained on only normal data, the model should be incapable of generating an anomalous output sequence. Anomalies are therefore detected by reconstructing a subsequence/sliding window from the test data and comparing it to the actual values, which generates a "reconstruction" error. Anomalies are generally flagged when the reconstruction probability falls below a specified threshold.

Most deep learning methods including generative models such as autoencoders (AEs), variational autoencoders (VAEs), generative adversarial networks (GANs), and transformers use reconstruction error as the anomaly score [29, 33]. Although these models are different in their architectures, training, and objective functions, most approaches using one of these models calculate anomaly scores as reconstruction errors. Note that in multivariable time series with multiscale spatiotemporal dependencies deciding what exactly constitutes "similarity" may be difficult. Fortunately, analytically defining similarity is usually not necessary and the anomaly score is often related to the loss function of the model. While reconstruction-based AD methods are fairly intuitive and quite widely used, they can be plagued with difficulties such as computational cost for data reconstruction, mode collapse, nonconvergence, and instability [33, 50].

3.1.2.3 Distance/dissimilarity

Distance-based models are based on a similarity metric that flags anomalies if their distance from normal is past a certain threshold. Clustering is an unsupervised machine learning model that is effective for grouping data and detecting anomalies; it involves mapping the time series data into a multidimensional space where the data are grouped near centroids based on feature similarity. Anomalies are then detected if they are far from existing clusters or have low probability of belonging to a cluster. Examples of clustering methods include K-means algorithm, one-class support vector machine (OCSVM), and Gaussian mixture model (GMM) [33]. More sophisticated machine learning methods such as Dynamic Time Warping provide a more complex comparison of temporal sequences (or subsequences, usually determined with a sliding window of fixed length) by allowing nonlinear alignment between sequences that are locally out of phase [11, 45]. Clustering methods are currently the benchmarks for anomaly detection in time series [39] and some recent studies demonstrate that many advanced algorithms do not deliver improved performance on basic univariate time series in comparison to more traditional methods and may in fact result in inferior performance [47]. The performance of clustering methods is generally degraded on complex high-dimensional datasets [33, 43], however. In such cases, methods to reduce data dimensionality such as expert opinion or various feature selection and extraction techniques such as deep autoencoders, principal component analysis (PCA) [51], and multidimensional scaling (MDS) [52, 53] can be used, in effect utilizing a hybrid approach using deep learning for dimensionality reduction and clustering methods for anomaly detection.

3.2 Capturing temporal context

The history of a sequence contains significant information regarding its potential future behavior and most deep learning approaches in some way depend on modeling the temporal dimension in order to explicitly capture the past during reconstruction or prediction.

3.2.1 Input

Input shape is essential to capturing temporal context and can take the form of individual (multidimensional) points or windows, which consist of a subsequence that contains some portion of the historical information. The width of the window is usually predetermined and can be based on the known or estimated/expected characteristics of the dataset and the presumed underlying process. Windows can be advanced by some number of data points (window step) and used in order (sliding windows) or they can be shuffled and entered out of order depending on the application and dataset. To specifically address the challenge of comparing subsequences rather than points, many models use representations of subsequences instead of the raw data. A sliding window decomposition/extraction is usually performed in the preprocessing stage after other operations such as missing value imputation, changing the sampling rate, or data normalization, have been completed [38, 39].

3.2.2 Temporal modeling

Several approaches to modeling temporal context and dependencies are commonly used in deep learning models. These essentially provide ways of organizing "memory," which amounts to in some way utilizing appropriately weighted previously encountered data in order to generate current or future output. These approaches may constitute the model architecture itself or they may be utilized at the midlevel of network dynamics and can be combined with various more basic activation functions as well as higher-level deep learning architectures.

3.2.2.1 Recurrent neural networks

Recurrent neural networks (RNNs) are a class of neural networks specifically designed for modeling sequential data and are well suited for capturing temporal and long-term dependencies. Unlike traditional feedforward neural networks, RNNs have a recurrent connection that allows information to be looped back and processed at each time step. This enables RNNs to retain memory of previous time steps that they can use to inform predictions at subsequent time steps. The hidden state of the RNN serves as an internal representation that evolves over time, capturing the context and history of the time series [52, 53]. A major shortcoming of RNNs is their instability due to the vanishing or exploding gradient problem, where the learning gradient becomes extreme as the network becomes deeper [54]. The earliest and most widely used RNN modification designed to address this problem is the long short-term memory (LSTM) unit [55], which avoids the problem by controlling retained information through a memory cell and input, output, and forget gates. If the LSTM unit detects an important feature from an earlier input sequence, it can carry this information over an extended distance, sometimes up to thousands of steps [39, 55]. A simpler and more computationally efficient but similarly effective modification is the gated recurrent unit (GRU) [56], which modulates the flow of information inside the unit but without a separate memory cell [52, 53]. Both LSTM and GRU cells are able to learn long-term dependencies by determining the number of weighted previous states to keep or forget at each time step. Both types of cells have been used with success in time series anomaly detection [39].

3.2.2.2 Convolutional neural networks

Convolutional neural networks (CNNs) have traditionally been used for image analysis but can be adapted to time series data and in some applications demonstrate better performance than RNNs [57–59], which still remain the most commonly used approach to temporal modeling. CNNs treat time series data as a one-dimensional (1D) array rather than a sequence and employ convolutional operations to capture local patterns and dependencies within the data. By applying one-dimensional convolutions, CNNs can learn hierarchical representation of time series and automatically extract relevant features at different timescales. Pooling layers can be added to downsample the output and reduce dimensionality. The learned features can then be fed into fully connected layers for classification or regression tasks [29, 33]. Anomaly detection can be performed by training the model on a normal dataset and then computing prediction or reconstruction error during inference. CNNs can be used for multidimensional time series as well as for real-time detection of anomalies; however, the extensive computational demands of CNNs make them less efficient for real-time monitoring. A specific approach for time series, the temporal convolutional network (TCN) [60], uses one-way convolutions in order to maintain temporally ordered/causal relations in the data. The TCN can generate sequences of any length and employs dilated convolutions, where the receptive field of the convolutional filters expands exponentially, which allow the model to capture long-range dependencies in time series.

Convolutional neural networks and RNNs can be combined in the same network architecture in order to capture spatial and temporal dependencies distinctly but simultaneously. In such models, CNNs are usually used to capture local patterns and features, with the output then fed into a RNN (either LSTM or GRU) that then processes the features extracted from the CNN in a sequential manner. The RNN performs the task of modeling the temporal context and relationships between the features extracted by the CNN [34, 61].

3.2.2.3 Attention

The attention mechanism, initially popularized in the context of natural language processing [62], has been extended to handle general time series data and provides a recent alternative to both RNN and CNN models. In the context of time series analysis, the attention mechanism allows the model to focus on specific temporal segments or patterns within the input sequence that are most relevant for making predictions. Using attention, the model can dynamically weigh the contribution of each time step based on its relevance, rather than relying solely on a fixed window or fixed-size context. Attention mechanisms typically involve a scoring mechanism that calculates the relevance or attention weight for each time step, followed by a weighted combination of the time step representations to produce a context vector that is then used for making predictions or further processing. Attention mechanisms in time series analysis have shown effectiveness in tasks such as sequence classification, forecasting, and anomaly detection, as they enable models to selectively attend to relevant temporal information while disregarding irrelevant or noisy segments of the time series [63–65].

3.2.2.4 Graph neural networks

Graph neural networks (GNNs) for time series data combine the power of graph structures with the ability to model temporal dependencies. GNNs enable the representation and analysis of time series data as graphs, where each data point is a node and the temporal relationships between them are represented as edges [66]. By incorporating recurrent or convolutional mechanisms, GNNs can capture the dynamics and patterns of time series data within the graph framework. These models leverage information from neighboring nodes and the temporal context to make predictions or perform tasks such as forecasting, classification, or anomaly detection. GNNs for time series offer a flexible and effective approach for handling complex temporal dependencies while leveraging the benefits of graph representations, enabling improved understanding and analysis of time-varying data [66]. GNN ideas have been implemented in graph convolutional networks [67] and graph attention networks [68] and are a promising future direction.

3.3 Summary

Deep learning models can automatically learn representations from raw time series data and capture both local and global dependencies, making them well suited for anomaly detection tasks. One of the key advantages of using deep learning for anomaly detection in time series is the ability to handle high-dimensional and complex data. Deep learning models such as RNNs, CNNs, and their variants, have been successfully applied to capture temporal dependencies, spatial patterns, and multiscale features in time series data and can effectively extract meaningful representations from the input data and detect anomalies based on deviations from learned normal patterns. These medium-scale models can be combined and integrated into larger scale unsupervised architectures such as autoencoders and GANs, which have been widely employed to learn compact representations of normal patterns in data. By reconstructing the input data or generating synthetic data samples, these models can detect anomalies by measuring the reconstruction error or the divergence between the real and generated data. More recent ways of capturing the temporal context include attention and GNNs, and these have shown promise in anomaly detection in time series.

4. Generative adversarial models for anomaly detection

4.1 Motivation for the use of GANs in medical data

The key motivation for utilizing GANs for anomaly detection in medical time series is that they almost directly address many of the challenges specific to medical time series discussed above [1, 14, 16–21]. Since GANs can learn to recreate normal data patterns and detect anomalies as deviations from these patterns, they are optimized to operate in an unsupervised setting; they can easily be combined with a temporal model (e.g., RNN or Attention) to model complex temporal and multivariate dependencies due to their ability to capture intricate data structures and patterns, nonlinearities, and high-dimensional relationships and they can do so in high-dimensional datasets involving multiple variables recorded over time; they can generate synthetic realistic data that can supplement original medical datasets, which often suffer from lack of volume and diversity, thus aiding in the training of more robust and generalizable models; and they have shown impressive generalization capabilities, allowing them to learn representations that generalize well to unseen data, which are particularly valuable in anomaly detection in medical time series, where the ability to accurately detect anomalies in previously unseen or rare cases is crucial.

4.2 GAN overview

Generative adversarial networks are a class of deep learning models introduced by Goodfellow et al. in 2014 [69] designed to generate synthetic data that resemble the training dataset by learning and reproducing the distribution inherent in the data. A GAN consists of two parts, a generator ($\mathcal{G}$) and a discriminator ($\mathcal{D}$), which are usually implemented as neural networks. Both networks play a two-player minimax game, in which $\mathcal{G}$ tries to generate data that $\mathcal{D}$ cannot distinguish from the real training data, while $\mathcal{D}$ tries to correctly classify data as real (from the training data) or fake (from $\mathcal{G}$).

The generator's role is to produce synthetic data samples, such as images, texts, or audio, which mimic the distribution of the training data. The generator is a mapping function that projects random noise or an input vector sampled from a prior distribution (which may be uniform or Gaussian) to the data space. The network gradually learns to transform the noise into output that is generated from the dataset distribution and is ideally indistinguishable from the training data. The discriminator, on the other hand, acts as a binary classifier. It is trained to distinguish between real data samples from the training set and synthetic samples generated by the generator. The discriminator takes both real and generated samples as inputs and outputs a probability score indicating the likelihood that the input is real. The objective of the discriminator is to learn to accurately classify the data samples.

The generator and the discriminator are trained simultaneously. Initially, the generator produces poor quality outputs and the discriminator can easily tell the difference between real and fake. As training progresses, the generator becomes better at generating fake outputs that appear real, and the discriminator becomes better at distinguishing between the real and the fake. The training continues until a point at which the discriminator can no longer distinguish fake data from real data (Nash equilibrium). The optimization process uses backpropagation and an optimization algorithm (like stochastic gradient descent) to adjust the parameters of $\mathcal{G}$ and $\mathcal{D}$. The loss function used for the training is typically binary cross-entropy.

DOI: http://dx.doi.org/10.5772/intechopen.112582

In the context of anomaly detection, after the GAN is trained, new data instances can be fed to the trained generator, which then attempts to reconstruct or regenerate these instances. The regenerated samples are then compared to the original data. If the difference between the original and the regenerated sample surpasses a defined threshold, the data instance is flagged as an anomaly. The underlying assumption is that the GAN will be less successful in accurately reproducing instances that are significantly different from the distribution it was trained on, i.e., anomalous instances. This procedure allows for effective anomaly detection in an unsupervised setting [27, 29–33].

The objective function summarizing this process is:

$$Min_{\mathcal{G}}Max_{\mathcal{D}}f(\mathcal{D},\mathcal{G}) = \mathbb{E}_{x\sim p_{data}(x)}[log\,(\mathcal{D}(x))] + \mathbb{E}_{z\sim p_z(z)}[log\,(1-\mathcal{D}(\mathcal{G}(z)))] \quad (1)$$

Where $\mathbb{E}_x$ and $\mathbb{E}_z$ are the expectation values over real and random data samples, respectively, $\mathcal{D}(\mathrm{x})$ is the probability estimate of the discriminator $\mathcal{D}$ if x is real, $\mathcal{G}(\mathrm{z})$ is the output of the generator $\mathcal{G}$ for a given vector z as input, and $\mathcal{D}(\mathcal{G}(\mathrm{z}))$ is the probability estimate of the discriminator $\mathcal{D}$ that the fake generated sample $\mathcal{G}(\mathrm{z})$ is real.

Generative adversarial networks have achieved remarkable success in various domains, including image synthesis, video generation, text generation, and more and have been applied to additional tasks such as image-to-image translation, style transfer, super-resolution, and data augmentation. The model structure described above is considered to be “vanilla” GAN. Since it is the most basic, the GAN model has been modified and extended to address specific challenges. For example, Wasserstein GAN [70] was developed to address unstable optimization and mode collapse; it shares the same minmax training procedure with the original GAN model but adjusts the loss function to minimize the Wasserstein distance between the real and fake data distributions. Conditional GAN (C-GAN) [71] adds conditioning information to both the generator and discriminator, which allows the model to generate data based on the chosen set of conditional parameters. Deep Convolutional GAN (DC-GAN) [72] uses convolutional layers in both generator and discriminator in order to generate more realistic images. Self-attention GAN (SA-GAN) [73] adds the self-attention mechanism to the convolutional GAN in order to model long-range, multilevel dependencies across image regions in order to avoid using only spatially local properties for generating high-resolution images. In Bidirectional GAN (Bi-GAN) [74], an encoder is added to the generator and discriminator in order to map the data into a latent space, enabling learning of a bidirectional mapping between the data space and the latent space. Finally, Cycle-consistent GAN (CycleGAN) involves training two GANs simultaneously using cycle consistency loss in order to encourage learned mappings in both directions and optimize image-to-image translation [75].

4.3 GANs for anomaly detection

Generative adversarial networks have proven to be effective as an unsupervised anomaly detection technique and have overcome significant challenges that are common to medical datasets such as a lack of adequately labeled datasets, dearth of anomalous data, and unbalanced datasets [28–31]. The following is a brief summary of the major models, which constitute the core models that have been adapted and modified for different applications.

4.3.1 AnoGAN

The AnoGAN model [76] was one of the earliest uses of GANs for anomaly detection. It uses a deep convolutional GAN (DC-GAN) architecture (see above) trained on normal data. Once the model is trained, anomaly scoring for a new instance is performed by calculating the anomaly score as a discrepancy between the instance and its reconstructed version obtained from the latent (random) space of the GAN. To reconstruct an instance, AnoGAN employs an optimization process that finds the closest point to it in the latent space by minimizing the difference between the reconstructed output and the original input using gradient descent and updating the latent code iteratively until convergence. The anomaly score is then calculated based on the difference between the original instance and the reconstructed output. A problem with this approach is that the GAN only implicitly models the data distribution and the optimization procedure for recovering the latent representation of a given sample is computationally costly and not practical for large datasets [29, 31]. The same authors followed up [77] with a modified model based on Wasserstein GAN, fast unsupervised anomaly detection with generative adversarial networks (f-AnoGAN), which substantially sped up the process of mapping to the latent space by moving from an iterative gradient descent approach to a learned mapping, which made the model more suitable for real-time anomaly detection.

4.3.2 Efficient GAN (E-GAN)

Efficient GAN [78] is based on AnoGAN, but it uses the bidirectional GAN (Bi-GAN) rather than a DC-GAN and incorporates an encoder into the architecture in order to alleviate the computational complexity associated with inference. Here, the discriminator separates two joint distributions: the given sample and the corresponding latent space (the output of the encoder) versus the original latent space and its generated synthetic sample (the output of the generator). The encoder acts as a regularization mechanism, helping to mitigate mode collapse and stabilizing the training process as well as resulting in significantly improved efficiency in detecting anomalies.

4.3.3 GANomaly

GANomaly [79] represents an addition of an autoencoder to AnoGAN in order to learn both the image and latent representations jointly. Here, the generator is constructed of an encoder and decoder, with an additional encoder that takes the output from the generated sample space and maps it back to a latent space. The discriminator as usual compares the generated samples to the original data. The total training loss then consists of the reconstructed loss in the latent space, the reconstructed loss in the sample space, and the adversarial loss in the sample space. The anomaly score is based on the encoder loss [31]. An important aspect of this model is that the space used for comparison is from the original sample space rather than depending on random sampling from a latent space as in other GAN-based models. This results in a model with high detection accuracy that has been adapted extensively to specific applications. Skip-GANomaly, for example, adds skip connections between the encoder and generator, leading to improved reconstruction accuracy and thus more accurate anomaly detection [29].

4.4 GANs for anomaly detection in time series

Generative adversarial networks offer a powerful framework for modeling and generating complex data distributions, making them well suited for capturing the intricate patterns and dynamics present in time series data. Several types of GAN architectures have been adapted for anomaly detection in time series. The following is a selected review of several important models.

4.4.1 TAnoGAN

Time Series Anomaly Detection with Generative Adversarial Networks (TAnoGAN) [80] is the simplest GAN-based model adapted specifically for time series. It is similar to the AnoGAN model, except that rather than using deep CNNs, both the generator and the discriminator are composed of LSTM layers in order to model temporality. A sliding window inputs real subsequences and the generator outputs simulated sequences of the same length, which are then compared by the discriminator using pointwise distance. A shortcoming of this model is that it is incapable of dealing with multivariate time series.

4.4.2 MAD-GAN

Multivariate Anomaly Detection GAN (MAD-GAN) [81] was designed specifically to address anomaly detection in multivariate time series. MAD-GAN also consists of LSTM layers in the generator and discriminator, and its training is similar to that of TAnoGAN. The model generates a residual loss based on the idea that the generator implicitly models the data distribution by learning to map it back into its latent space. During inference, this residual loss is combined with the standard discrimination loss to determine whether the sample is abnormal or not.

4.4.3 TadGAN

Time series anomaly detection using generative adversarial network (TadGAN) [82] also uses the LSTM structure in its generator and discriminator, but introduces an encoder in order to generate the latent space from which the generator produces synthetic data samples (rather than from a random latent space as usual). There are two discriminators, one for the samples and one for the latent space. The model uses Wasserstein distance for the discriminator loss and cycle consistency loss for the reconstruction error, which are computed through a combination of pointwise difference, area difference, and dynamic time warping. Anomalies are detected through a combination of reconstruction and discriminator errors.

4.4.4 Beat-GAN

Beat-GAN [59] was specifically developed for ECG anomaly detection. It utilizes an encoder-decoder architecture as the generator, and both the encoder and decoder are CNNs rather than LSTMs. They deal with temporality by utilizing a one-dimensional filter sliding along the temporal dimension. An interesting contribution that this model uses in order to deal with the inherent nonstationarity of their data (heart rate variability) is a modified form of time warping, where data are imputed during decelerations and removed during accelerations in order to generate a steady "beat." The model functions essentially like an autoencoder, but uses the discriminator for regularization to improve stability. During inference, anomalies are detected by a combination of pairwise reconstruction error and discriminator error.

5. GANs in medicine

5.1 GANs in medical imaging

Due to their capacity to generate realistic images, it is not surprising that the most common use of GANs in medical imaging has been primarily in data synthesis [83–85]. While some of these uses do not directly involve anomaly detection, given that they are in some way involved in detecting and diagnosing specific disease or behavioral states, they are at least obliquely related [21, 28]. The following are some of the most common current applications.

5.1.1 Data augmentation

As previously discussed, scarcity of labeled data represents one of the main limitations to the application of deep learning in medicine [86, 87]. Medical datasets are often imbalanced and lack diversity, which can lead to biased or poor-performing models. GANs have been shown to be able to generate synthetic medical data, helping to augment existing datasets, rectify imbalance, increase diversity, and improve the performance of machine learning models. A key advantage of being able to generate synthetic data with the same statistical characteristics as the original data but without personal health information is the ability to widely share and analyze data without the risk of violating patient privacy, which is often a barrier to producing large public datasets.

From an anomaly detection perspective, data synthesis can be used to turn an unsupervised or semi-supervised anomaly problem into a supervised binary (or multiclass) classification problem: GANs are used to generate synthetic data that statistically resemble anomalies, which can then be used for balanced classification training in a supervised manner. This is, in fact, the most common current application of GANs in anomaly detection [21, 29, 30] and has been applied to images [83, 84] as well as time series such as ECG [88] and EEG [89]. While this approach to anomaly recognition is more straightforward than classical anomaly detection, it assumes that the known anomalous data cover the entire distribution of possible anomalies, which may not be the case and could result in excellent classification of known anomalies but possible misclassification of unknown ones [39, 85].

5.1.2 Image-to-image translation

A powerful application specific to GANs is their ability for image-to-image translation such as converting MRI images to computed tomography (CT) images or vice versa, enhancing the quality of medical images, or generating angiography images from MRI images. There are multiple potential benefits to this, including decreased need for multimodal studies, reduced acquisition time and radiation exposure, and increased availability of appropriate imaging in cases where there is limited access to multimodal imaging, e.g., in a clinical scenario where a certain imaging modality (such as MRI) might be optimal for diagnosis or treatment planning but only another modality (such as CT) is available to the patient. The predominant type of GAN used for this application is CycleGAN [21, 90–93].

5.1.3 Super-resolution/denoising

Generative adversarial networks can also be used to increase the resolution of medical images, known as super-resolution, and to reduce noise and artifacts. Since the quality of medical images is often degraded by various factors such as hardware limitation or patient movement, this can be extremely helpful for discernment of fine details that can be critical for correct diagnosis. Super-resolution and noise reduction are types of image-to-image translation that involve converting low-resolution images into high-resolution images by imputing data. This is especially interesting in cases where the imaging modality is intrinsically low-resolution, such as positron emission tomography (PET) [94–99].

5.1.4 Image segmentation

Image segmentation is important for measuring and visualizing anatomical structures, delineating pathological regions, and for surgical planning and image-guided interventions. The process of applying GANs to image segmentation is slightly different than that of applying vanilla GAN: the generator now aims to create an image where each pixel corresponds to a particular class label and the discriminator attempts to differentiate between the ground-truth segmentation (real) and the generator's segmentation (synthetic). This is again a type of image-to-image translation for which GANs have been used successfully [100–102].

5.1.5 Disease progression modeling

Disease progression modeling involves predicting how a disease will develop in a patient over time, which can allow for early intervention, optimized treatment plans, and better patient outcomes. In the context of disease progression modeling, the generator could be conditioned on a particular disease stage or on past medical history in order to generate synthetic data predicting what could potentially happen at the next disease stage or time point. Different models or approaches might be used to handle different kinds of data (continuous, discrete, etc.) and different diseases. GANs have been successfully applied to tumor growth [103] and Alzheimer's disease prediction [104].

5.1.6 Brain decoding

Brain decoding involves using machine learning algorithms to map patterns of brain activity (measured via EEG or fMRI) to mental states or processes. For example, using visual image reconstruction decoding researchers have been able to reproduce images a person is viewing directly from brain activity [105–107]. This process is often referred to as "mind reading" or "brain-to-image reconstruction". The most direct clinical applicability of this technique at this time is in brain-computer interfaces, which would potentially allow disabled and paralyzed individuals to communicate and control external devices more easily. The technique also offers the potential for significant advancements in the understanding of the biology of consciousness and mind-body medicine.

5.2 GANs in medical time series: ECG

5.2.1 ECG overview

Electrocardiography (ECG or EKG) is a diagnostic tool that records the electrical activity of the heart over a period of time through electrodes placed on the skin, typically in a standard 12-lead setup for a clinical ECG. The ECG waveform represents the electrical depolarization and repolarization of the cardiac muscle during each heartbeat and can provide a large amount of valuable information such as the heart rate, rhythm, and the size and position of the chambers. It can also show evidence of damage to the cardiac muscle (ischemia or infarction), effects of drugs or devices (such as a pacemaker), and other types of heart disease or conditions (e.g., pericarditis, electrolyte imbalances). A significant advantage of ECG is that it is noninvasive, inexpensive, and relatively quick to perform. However, it requires expert interpretation, and while it is highly valuable, it may not provide a definitive diagnosis on its own and may need to be combined with other studies [108, 109].

5.2.2 GANs in ECG anomaly detection

As discussed above, Beat-GAN was specifically designed to detect anomalies in ECG. It outperformed other anomaly detection methods (including OCSVM) and achieved high accuracy and fast inference. It was also to some extent interpretable since it was able to pinpoint anomalies in sample space. The model has also been applied successfully to time series in other domains [59]. Shin et al. [110] deployed the AnoGAN architecture, but modified it with extensive data preprocessing as well as dimensionality reduction with t-distributed stochastic neighbor embedding (t-SNE), which they utilized to generate an experimentally determined objective decision boundary that could effectively differentiate between normal signal and arrhythmia based on anomaly score.

Li et al. [111] proposed Single-Lead convolutional generative adversarial network (SLC-GAN) for automated myocardial infarction (MI) in single-lead ECG. The model involves a GAN with multiple convolutional layers (DC-GAN) in the generator and discriminator with an added CNN classifier for MI detection. The GAN portion learns to generate synthetic ECG data, which are then used to augment the volume of the training data for the classifier. The model achieved excellent classification accuracy and provides a good example of using synthetic data in order to change the problem parameters from unsupervised to supervised and improve performance. Xia et al. [112] extended this idea by applying a transformer model in the generator with a CNN discriminator. This model was then used to generate synthetic data, which were used to augment training of a classifier that combined a CNN-based feature extraction block followed by a bidirectional LSTM (Bi-LSTM) architecture. The overall model achieved superior performance and demonstrated improved classification than models that did not use added synthetic data. A similar conclusion was reached by Rath et al. [113], who tested several machine and deep learning methods and found that the best performance on ECG classification was achieved by a GAN-LSTM ensemble model.

Qin et al. [114] proposed ECG-ADGAN, a semi-supervised model that incorporates a Bi-LSTM network in addition to multiple 1D convolutional layers into the GAN generator in order to preserve temporal patterns of the ECG signal. Training takes place in two stages, with stage I following normal GAN training to Nash equilibrium between the generator and discriminator and stage II freezing the generator and training the discriminator/classifier separately specifically for anomaly detection. The authors also utilized mini-batch training during stage I to improve convergence. The model demonstrated superior detection of unknown anomalies when compared to supervised learning methods.

Wang and colleagues [115] further extended this approach beyond binary classification (normal vs. abnormal) by incorporating GAN into a two-level hierarchical framework in order to not only detect but also classify different types of arrhythmias. The first level consists of a memory-augmented deep autoencoder with GAN (MadeGAN) designed to perform anomaly detection; the second level consists of a multibranching deep CNN architecture utilizing transfer learning to allow classification of different types of heart disease given the fundamental imbalance in the training dataset. This framework was able to effectively capture disease-altered features of ECG signals and accurately predict and classify heart disease with better performance compared to existing methods.

5.3 GANs in medical time series: EEG

5.3.1 EEG overview

An electroencephalogram (EEG) is a neuroimaging technique used to record the electrical activity of the brain. It is carried out using multiple electrodes placed on the scalp according to a standardized placement system, usually the 10–20 system. These electrodes measure voltage fluctuations resulting from ionic current flows within neural populations in the brain. The resulting traces, EEG waves, represent the summation of postsynaptic potentials (PSPs) from a large number of neurons, specifically from cortical pyramidal neurons, detected as fluctuations in voltage over time [116].

EEG waves are characterized by their frequency (measured in Hertz), amplitude (measured in microvolts), and waveform morphology. They are typically categorized into bandwidths: delta (0.5–4 Hz), theta (4–8 Hz), alpha (8–12 Hz), beta (12–30 Hz), and gamma (30–100 Hz), each of which may correspond to different states of brain activity or consciousness. For instance, alpha waves are typically associated with relaxed, closed-eye states, while beta waves are associated with active thinking, attentional focus, or rapid eye movement (REM) sleep [116].

EEG can be used for the detection and study of various neurological and psychiatric conditions such as epilepsy, sleep disorders, encephalopathies, and even cognitive processes. The main advantage of EEG in cognitive imaging is its high temporal resolution, which allows for the study of fast-dynamic processes within the brain. Its spatial resolution is limited, however, and it is less effective for capturing activity occurring deep within the brain. Anomaly detection in EEG traces involves all the challenges discussed above: multidimensional time series with complex dependencies, highly complex and nonstationary normal behavior, and rarity and extreme heterogeneity of disease patterns, from normal brain aging to active seizures.

5.3.2 GANs in EEG epilepsy detection

Epilepsy is a chronic neurological disorder marked by recurrent seizures, which are symptoms of abnormal excessive or synchronous neuronal activity in the brain. Seizures can vary greatly in their presentation, from minor sensory disturbances or momentary lapses in consciousness ("absence seizures") to full-body convulsions ("grand mal seizure"). EEG is crucial in diagnosing and localizing the disorder, as well as in detecting seizures when they occur. Seizures are characterized by a variety of EEG patterns and frequencies such as spike-and-wave discharges (70 ms waveforms often followed by a slow wave, with specificity based on the frequency band), sharp waves (typically seen in focal seizures), polyspikes (typically seen in generalized seizures), focal slowing (which helps with localization and can be seen before, during, or after focal seizures), or generalized paroxysmal fast activity (rapid continuous spiking typically seen in severe diseases such as Lennox-Gastaut syndrome). Seizure detection and monitoring has an important role in diagnosis, improving quality of life, and general understanding of the disease. For example, alerting a patient about an impending seizure might allow them to take appropriate safety precautions or break-through medications for seizure control [117–121]. In this context, automatic seizure detection or prediction essentially consists of a binary classification task between the ictal (seizure) or pre-ictal and nonictal (nonseizure) EEG patterns.

A number of machine and deep learning approaches have been applied to epileptic seizure detection and prediction, but these predominantly apply feature extraction such as wavelets or independent component analysis (ICA) followed by a classifier such as random forest or support vector machine (SVM) [122]. One of the only models to directly apply a semi-supervised GAN model to seizure detection was introduced by You et al. [119], who modified the AnoGAN model (DC-GAN architecture) for seizure detection in a behind-the-ear EEG two-channel signal. The EEG signal channels were filtered and transformed into spectrogram images, which were combined to form a virtual channel image that was fed to the network for training. The GAN was trained on normal data and anomalies were detected using a combination of residual and discriminative loss. The authors noted that the addition of a Gram matrix of the feature maps from each convolutional layer was shown to improve performance. The model demonstrated a 96.3% sensitivity for automated seizure detection.

Zhu and Shoaran [123] utilized an unsupervised adversarial model to map power spectrum features from intracranial EEG recordings into a subject-invariant feature space via domain transfer learning. The model consisted of an encoder and decoder functioning as a generator for both the source (labeled data) and target (unlabeled or new data) domains, with a discriminator designed to try to differentiate between the domains. A discriminative model was then trained on the resultant subject-invariant features to generate predictions about the target patients. The model demonstrated improved performance compared to the more conventional subject-specific approach, allowing for better generalization.

In contrast, Truong et al. [124] used a GAN to extract features from the EEG data that could then undergo binary classification. The generator was trained to synthesize realistic short-time Fourier transform images from a noise vector that were then passed through the discriminator. After training, the discriminator was able to collect and flatten features that could be used with any generic classifier, in this case two fully connected layers. The model again consisted of a modified DC-GAN architecture that was trained in an unsupervised manner where information regarding seizure onset was disregarded.

While there are very few applications of GANs directly to seizure detection, most applications use GANs in order to produce balanced training datasets with generated synthetic data, again effectively changing the problem from an unsupervised to a supervised one. One of the first applications of GANs to EEG data generation was proposed by Pascual et al. [125] who used their conditional GAN model to generate

DOI: http://dx.doi.org/10.5772/intechopen.112582

synthetic ictal signals conditioned on interictal data from individual patients. The model consisted of a convolutional autoencoder as the generator, in this case translating the latent code into an ictal sample rather than restoring the original sample. The discriminator had the same architecture as the encoder of the generator and was trained to distinguish between real and fake ictal signals. The inclusion of synthetic samples resulted in improved classification performance compared to training on only real samples. The additional benefit of the synthetic procedure, as the authors point out, was deidentification of the original data and significant improvement in data privacy. Multiple additional groups have applied similar data augmentation approaches with various modifications, including different feature extraction methods, different generator and discriminator architectures, different loss functions, utilization of LSTM/GRU cells or attention instead of CNNs, and the application of different classifiers [126–143].

6. Conclusion: challenges and future directions

Although GANs have shown significant promise in anomaly detection in medical time series, serious challenges remain, such as (notorious) training instability, lack of clear evaluation metrics for generated data, limited interpretability, inability to explicitly model causal relationships, question of preservation of temporal dynamics, privacy concerns, and ethical considerations in the use of generated data. For example, evaluating the performance of GAN-based anomaly detection in time series is nontrivial since traditional evaluation metrics such as precision, recall, or F1-score or even more advanced techniques such as maximum mean discrepancy or Fréchet inception distance [144, 145] may be unable to fully capture the highly complex characteristics of the data. Additional issues such as fairness in AI models also need to be considered to ensure that algorithmic decisions do not discriminate against certain demographic groups.

Future research efforts should focus on refining training techniques, incorporating domain knowledge, and developing hybrid approaches to enhance the performance and applicability of GANs in anomaly detection in time series data. Since GANs were originally designed to generate realistic images, their use is still predominantly focused on the generation of synthetic data. This is, of course, extremely useful as it can help resolve issues of unbalanced datasets and lack of anomalous and labeled data and transform unsupervised or semi-supervised approaches to anomaly detection into more manageable supervised problems. Direct anomaly detection with GANs is becoming more common, however, as different research groups realize that applying generative deep learning to data directly is feasible and provides significant advantages (and since generation of synthetic data inherently biases the network away from correct classification or detection of unknown or unusual anomalies). While GANs have been applied successfully, at least initially, in anomaly detection in time series such as ECG and EEG, there are many other areas where they could be extremely useful, such as rs-fMRI, ICU physiological data, or even monitoring of medical records for the possibility of subtle signs of chronic disease. In these areas, the application would leverage the model's ability to learn the underlying distribution, and modifications such as adding or embedding different modules (such as autoencoders) into the adversarial structure and/or including recurrence, attention, or graph structures could serve to better model the long-range spatiotemporal dynamics and dependencies in the data. Additional potential areas of future research include building patient-specific models, improving generality through transfer learning, and real-time applications in the monitoring of healthcare data. Finally, there is a need for more research on the practical deployment of GANs in clinical settings, which involves not only technical considerations but also evaluation in terms of clinical outcomes, cost-effectiveness, user experience, and ethical, legal, and societal implications.

Author details

Miloš Cekić
University of California, Los Angeles, Los Angeles, California, USA

*Address all correspondence to: mcekic@mednet.ucla.edu

References

[1] Ching T, Himmelstein DS, Beaulieu-Jones BK, Kalinin AA, Do BT, Way GP, et al. Opportunities and obstacles for deep learning in biology and medicine. Journal of the Royal Society Interface. 2018;**15**:2017038. DOI: 10.1098/rsif.2017.0387

[2] Roy S, Meena T, Lim SJ. Demystifying supervised learning in healthcare 4.0: A new reality of transforming diagnostic medicine. Diagnostics. 2022;**12**:2549. DOI: 10.3390/diagnostics12102549

[3] Kaushik S, Choudhury A, Sheron PK, Dasgupta N, Natarajan S, Pickett LA, et al. AI in healthcare: Time-series forecasting using statistical, neural, and ensemble architectures. Frontiers in Big Data. 2020;**3**:4. DOI: 10.3389/fdata.2020.00004

[4] Wang WK, Chen I, Hershkovich L, Yang J, Shetty A, Singh G, et al. A systematic review of time series classification techniques used in biomedical applications. Sensors. 2022;**22**(20):8016. DOI: 10.3390/s22208016

[5] Kline A, Wang H, Li Y, Dennis S, Hutch M, Xu Z, et al. Multimodal machine learning in precision health: A scoping review. NPJ Digital Medicine. 2022;**5**:171. DOI: 10.1038/s41746-022-00712-8

[6] Fernando T, Gammulle H, Denman S, Sridharan S, Fookes C. Deep learning for medical anomaly detection—A survey. ACM Computing Surveys. 2021;**54**(7):141. DOI: 10.1145/3464423

[7] Tschuchnig ME, Gadermayr M. Anomaly detection in medical imaging—A mini review. arXiv. arXiv preprint arXiv:2108.11986. 2021. DOI: 10.48550/arXiv.2108.11986

[8] Samariya D, Ma J. Anomaly detection on health data. In: Traina A, Wang H, Zhang Y, Siuly S, Zhou R, Chen L, editors. Health Information Science (HIS 2022). Cham, Switzerland: Springer Nature; 2022; LNCS, (13705):34-41. DOI: 10.1007/978-3-031-20627-6_4

[9] Chalapathy R, Chawla S. Deep learning for anomaly detection: A survey. arXiv. arXiv preprint arXiv:1901.03407. 2019. DOI: 10.48550/arXiv.1901.03407

[10] Pang G, Shen C, Cao L, Van Den Hengel A. Deep learning for anomaly detection: A review. ACM Computing Surveys. 2022;**54**(2):1-38. DOI: 10.1145/3439950

[11] Li G, Jung JJ. Deep learning for anomaly detection in multivariate time series: Approaches, applications, and challenges. Information Fusion. 2023;**91**:93-102. DOI: 10.1016/j.inffus.2022.10.008

[12] Gui J, Sun Z, Wen Y, Tao D, Ye J. A review on generative adversarial networks: Algorithms, theory, and applications. IEEE Transactions on Knowledge and Data Engineering. 2023;**35**(4):3313-3332. DOI: 10.1109/TKDE.2021.3130191

[13] Aggarwal A, Mittal M, Battineni G. Generative adversarial network: An overview of theory and applications. International Journal of Information Management, Data Insights. 2021;**1**:100004. DOI: 10.1016/j.jjimei.2020.100004

[14] Dash A, Ye J, Wang G. A review of generative adversarial networks (GANs) and its applications in a wide variety of disciplines–from medical to remote sensing. International Journal of Applied Earth Observation and Geoinformation.

2021;**108**:102734. DOI: 10.48550/arXiv.2110.01442

[15] Jabbar A, Li X, Omar B. A survey on generative adversarial networks: Variants, applications, and training. ACM Computing Surveys (CSUR). 2021; **54**(8):1-49. DOI: 10.1145/3463475

[16] Yi X, Walia E, Babyn P. Generative adversarial network in medical imaging: A review. Medical Image Analysis. 2019; **2019**(58):101552. DOI: 10.1016/j.media.2019.101552

[17] Koshino K, Werner RA, Pomper MG, Bundschuh RA, Toriumi F, Higuchi T, et al. Narrative review of generative adversarial networks in medical and molecular imaging. The Annals of Translational Medicine. 2021;**9**(9):821. DOI: 10.21037/atm-20-6325

[18] Lan L, You L, Zhang Z, Fan Z, Zhao W, Zeng N, et al. Generative adversarial networks and its applications in biomedical informatics. Frontiers in Public Health. 2020;**8**:164. DOI: 10.3389/fpubh.2020.00164

[19] Kazeminia S, Baur C, Kuijper A, van Ginneken B, Navab N, Albarqouni S, et al. GANs for medical image analysis. Artificial Intelligence in Medicine. 2020; **109**:101938. DOI: 10.1016/j.artmed.2020.101938

[20] Iqbal A, Sharif M, Yasmin M, Raza M, Aftab S. Generative adversarial networks and its applications in the biomedical image segmentation: A comprehensive survey. The International Journal of Multimedia Information Retrieval. 2022;**11**:333-368. DOI: 10.1007/s13735-022-00240-x

[21] Laino ME, Cancian P, Politi LS, Della Porta MG, Saba L, Savevski V. Generative adversarial networks in brain imaging: A narrative review. Journal of Imaging. 2022;**8**:83. DOI: 10.3390/jimaging8040083

[22] Soomro TA, Zheng L, Afifi AJ, Ali A, Soomro S, Yin M, et al. Image segmentation for MR brain tumor detection using machine learning: A review. IEEE Reviews in Biomedical Engineering. 2023;**16**:70-90. DOI: 10.1109/RBME.2022.3185292

[23] Krithika M, Suganthi K. Review of medical image synthesis using GAN techniques. ITM Web of Conferences. 2021;**37**:01005. DOI: 10.1051/itmconf/20213701005

[24] Ali H, Biswas R, Mohsen F, Shah U, Alamgir A, Mousa O, et al. The role of generative adversarial networks in brain MRI: A scoping review. Insights Into Imaging. 2022;**13**:98. DOI: 10.1186/s13244-022-01237-0

[25] Jeong JJ, Tariq A, Adejumo T, Trivedi H, Gichoya JW, Banerjee I. Systematic review of generative adversarial networks (GANs) for medical image classification and segmentation. Journal of Digital Imaging. 2022;**35**:137-152. DOI: 10.1007/s10278-021-00556-w

[26] Yahaya MSM, Teo J. Data augmentation using generative adversarial networks for images and biomarkers in medicine and neuroscience. Frontiers in Applied Mathematics and Statistics. 2023;**9**: 1162760. DOI: 10.3389/fams.2023.1162760

[27] Sabuhi M, Zhou M, Bezemer CP, Musilek P. Applications of generative adversarial models in anomaly detection: A systematic literature review. IEEE Access. 2021;**9**:161003-161029. DOI: 10.1109/ACCESS.2021.3131949

[28] Wang R, Bashyam V, Yang Z, Yu F, Tassopoulou V, Chitapalli SS, et al. Applications of generative adversarial networks in neuroimaging and clinical neuroscience. NeuroImage. 2023;**269**: 119898. DOI: 10.1016/j.neuroimage.2023.119898

[29] Li H, Li Y. Anomaly detection based on GAN: A survey. Applied Intelligence. 2023;**53**:8209-8231. DOI: 10.1007/s10489-022-03905-6

[30] Di Mattia F, Galeone P, De Simoni M, Ghelfi E. A survey on GANs for anomaly detection. arXiv. arXiv preprint arXiv:1906.11632. 2019. DOI: 10.48550/arXiv.1906.11632

[31] Esmaeili M, Toosi A, Roshanpoor A, Changizi V, Ghazisaeedi M, Rahmim A, et al. Generative adversarial networks for anomaly detection in biomedical imaging: A study on seven medical image datasets. IEEE Access. 2023;**11**: 17906. DOI: 10.1109/ACCESS.2023.3244741

[32] Chen X, Konukoglu E. Unsupervised abnormality detection in medical images with deep generative methods. In: Biomedical Image Synthesis and Simulation: Methods and Applications. London, UK; Academic Press; 2022. DOI: 10.1016/B978-0-12-824349-7.00022-0

[33] Choi K, Yi J, Park C, Yoon S. Deep learning for anomaly detection in time-series data: Review, analysis, and guidelines. IEEE Access. 2021;**9**:120043. DOI: 10.1109/ACCESS.2021.3107975

[34] Hamilton JD. Time Series Analysis. Princeton, NJ, USA: Princeton University Press; 2020. DOI: 10.1515/9780691218632

[35] Mills T. Applied Time Series Analysis: A Practical Guide to Modeling and Forecasting. London, UK: Academic Press; 2019. DOI: 10.1016/B978-0-12-813117-6.00001-6

[36] Shumway RH, Stoffer DS. Time Series Analysis and its Applications. 4th ed. New York, NY, USA: Springer; 2017. DOI: 10.1007/978-3-319-52452-8

[37] Chandola V, Banerjee A, Kumar V. Anomaly detection for discrete sequences: A survey. IEEE Transactions on Knowledge and Data Engineering. 2012;**24**(5):823-839. DOI: 10.1109/TKDE.2010.235

[38] Blazquez-Garcıa A, Conde A, Mori U, Lozano JA. A review on outlier/anomaly detection in time series data. ACM Computing Surveys. 2021;**54**(3): 1-33. DOI: 10.1145/3444690

[39] Darban AA, Webb GI, Pan S, Aggarwal CC, Salehi M. Deep learning for time series anomaly detection: A survey. arXiv. arXiv preprint arXiv:2211.05244. 2022. DOI: 10.48550/arXiv.2211.05244

[40] Brophy E, Wang Z, She Q, Ward T. Generative adversarial networks in time series: A systematic literature review. ACM Computing Surveys. 2023;**55**(10): 199. DOI: 10.1145.3559540

[41] Hawkins DM. Identification of Outliers. London, UK: Springer Netherlands; 1980. DOI: 10.1007/978-94-015-3994-4

[42] Ranga, Suri NNR, Murty N, Athithan MG. Outlier Detection: Techniques and Applications. New York, NY, USA: Springer; 2019. DOI: 10.1007/978-3-030-05127-3

[43] Aggarwal CC. Outlier Analysis. 2nd ed. New York, NY, USA: Springer; 2017. DOI: 10.1007/978-3-319-47578-3

[44] Munir M, Chattha MA, Dengel A, Ahmed S. A comparative analysis of traditional and deep learning-based anomaly detection methods for streaming data. In: 18th IEEE International Conference on Machine Learning and Applications (ICMLA). Los Alamitos, CA, USA: IEEE Computer Society Conference Publishing Services; 2019. pp. 561-566. DOI: 10.1109/ICMLA.2019.00105

[45] Ruff L, Kauffmann JR, Vandermeulen RA, Montavon G, Samek W, Kloft M, et al. A unifying review of deep and shallow anomaly detection. Proceedings of the IEEE. 2021; **109**(5):756-795. DOI: 10.1109/JPROC.2021.3052449

[46] Rewicki F, Denzler J, Niebling J. Is it worth it? Comparing six deep and classical methods for unsupervised anomaly detection in time series. Applied Sciences. 2023;**13**(3):1778. DOI: 10.3390/app13031778

[47] Audibert J, Michiardi P, Guyard F, Marti S, Zuluaga MA. Do deep neural networks contribute to multivariate time series anomaly detection? Pattern Recognition. 2022;**132**:108945. DOI: 10.1016/j.patcog.2022.108945

[48] Baid U et al. The RSNA-ASNR-MICCAI BraTS 2021 benchmark on brain tumor segmentation and radiogenomic classification. arXiv. arXiv preprint arXiv:2107.02314. 2020. DOI: 10.48550/arXiv.2107.02314

[49] Petzsche MRH et al. ISLES 2022: a multi-center magnetic resonance imaging stroke lesion segmentation dataset. Scientific Data. 2022;**9**:762. DOI: 10.1038/s41597-022-01875-5

[50] Hsu CY, Liu WC. Multiple time-series convolutional neural network for fault detection and diagnosis and empirical study in semiconductor manufacturing. Journal of Intelligent Manufacturing. 2020;**32**:1-14. DOI: 10.1007/s10845-020-01591-0

[51] Bao Y, Tang Z, Li H, Zhang Y. Computer vision and deep learning–based data anomaly detection method for structural health monitoring. Structural Health Monitoring. 2019; **18**(2):401-421. DOI: 10.1177/1475921718757405

[52] Tang Z, Chen Z, Bao Y, Li H. Convolutional neural network-based data anomaly detection method using multiple information for structural health monitoring. Structural Control and Health Monitoring. 2019;**26**(1): e2296. DOI: 10.1002/stc.2296

[53] Chung J, Gulcehre C, Cho KH, Bengio Y. Empirical evaluation of gated recurrent neural networks on sequence modeling. arXiv preprint arXiv: 1412.3555. 2014; arXiv. DOI: 10.48550/arXiv.1412.3555

[54] Bengio Y, Simard P, Frasconi P. Learning long-term dependencies with gradient descent is difficult. IEEE Transactions on Neural Networks. 1994; **5**(2):157-166. DOI: 10.1109/72.279181

[55] Hochreiter S, Schmidhuber J. Long short-term memory. Neural Computation. 1997;**9**(8):1735-1780. DOI: 10.1162/neco.1997.9.8

[56] Cho K, van Merrienboer B, Gulcehre C, Bahdanau D, Bougares F, Schwenk H, et al. Learning phrase representations using RNN encoder-decoder for statistical machine translation. arXiv. arXiv preprint arXiv: 1406.1078. 2014. DOI: 10.48550/arXiv.1406.1078

[57] Choi Y, Lim H, Choi H, Kim IJ. GAN-based anomaly detection and

localization of multivariate time series data for power plant. In: Proc. IEEE Int. Conf. Big Data Smart Comput. (BigComp). Los Alamitos, CA, USA: IEEE Computer Society Conference Publishing Services; 2020. pp. 71-74. DOI: 10.1109/BigComp48618.2020.00-97

[58] Wen T, Keyes R. Time series anomaly detection using convolutional neural networks and transfer learning. arXiv. arXiv preprint arXiv:1905.13628. 2019. DOI: 10.48550/arXiv.1905.13628

[59] Zhou B, Liu S, Hooi B, Cheng X, Ye J. BeatGAN: Anomalous rhythm detection using adversarially generated time series. In: Proc. 28th Int. Joint Conf. Artif. Intell. Menlo Park, CA, USA: AAAI Press; 2019. pp. 4433-4439. DOI: 10.24963/ijcai.2019/616

[60] Lea C, Vidal R, Reiter A, Hager GD. Temporal convolutional networks: A unified approach to action segmentation. arXiv. arXiv preprint arXiv:1611.05267. 2016. DOI: 10.48550/arXiv.1611.05267

[61] Mamandipoor B, Majd M, Sheikhalishahi S, Modena C, Osmani V. Monitoring and detecting faults in wastewater treatment plants using deep learning. Environmental Monitoring and Assessment. 2020;**192**(2):1-12. DOI: 10.1007/s10661-020-8064-1

[62] Vaswani A, Shazeer N, Parmar N, Uszkoreit J, Jones L, Gomez AN, et al. Attention is all you need. arXiv. arXiv preprint arXiv:1706.03762. 2017. DOI: 10.48550/arXiv.1706.03762

[63] Guo Y, Liao W, Wang Q, Yu L, Ji T, Li P. Multidimensional time series anomaly detection: A GRU-based Gaussian mixture variational autoencoder approach. In: Proceedings of the 10th Asian Conference on Machine Learning. Brookline, MA, USA: MLR Press/Microtome Publishing; 2018. pp. 97-112

[64] Lee TJ, Gottschlich J, Tatbul N, Metcalf E, Zdonik S. Greenhouse: A zero-positive machine learning system for time-series anomaly detection. arXiv. arXiv preprint arXiv:1801.03168. 2018. DOI: 10.48550/arXiv.1801.03168

[65] Lu Z, Lv W, Xie Z, Du B, Xiong G, Sun L, et al. Graph sequence neural network with an attention mechanism for traffic speed prediction. ACM Transactions on Intelligent Systems and Technology (TIST). 2022, 2022;**13**(2): 1-24. DOI: 10.1145/3470889

[66] Wu L, Cui P, Pei J, Zhao L, Song L. Graph neural networks. In: Wu L, Cui P, Pei J, Zhao L, editors. Graph Neural Networks: Foundations, Frontiers, and Applications. Singapore: Springer; 2022. DOI: 10.1007/978-981-16-6054-2_3

[67] Kipf TN, Welling M. Semi-supervised classification with graph convolutional networks. arXiv. arXiv preprint arXiv:1609.02907. 2016. DOI: 10.48550/arXiv.1609.02907

[68] Veličković P, Cucurull G, Casanova A, Romero A, Lio P, Bengio Y. Graph attention networks. arXiv. arXiv preprint arXiv:1701.10903. 2017. DOI: 10.48550/arXiv.1710.10903

[69] Goodfellow I, Pouget-Abadie J, Mirza M, Xu B, Warde-Farley D, Ozair S, et al. Generative adversarial nets. Neural Information Processing Systems. 2014;**2**:2672-2680. DOI: 10.3156/jsoft.29.5_177_2

[70] Arjovsky M, Chintala S, Bottou L. Wasserstein generative adversarial networks. Proceedings of the 34th international conference on machine learning. PMLR. 2017;**70**:214-223

[71] Mirza M, Osindero S. Conditional generative adversarial nets. arXiv. arXiv preprint arXiv:1411.1784. 2014. DOI: 10.48550/arXiv.1411.1784

[72] Radford A, Metz L, Chitala S. Unsupervised representation learning with deep convolutional generative adversarial networks. arXiv. arXiv preprint arXiv:1511.06434. 2015. DOI: 10.48550/arXiv.1511.06434

[73] Zhang H, Goodfellow I, Metaxas D, Odena A. Self-attention generative adversarial networks. arXiv. arXiv preprint arXiv:1805.08318. 2018. DOI: 10.48550/arXiv.1805.08318

[74] Donahue J, Krahenbuhl P, Darrell T. Adversarial feature learning. arXiv. arXiv preprint arXiv:1605.09782. 2016. DOI: 10.48550/arXiv.1605.09782

[75] Zhu JY, Park T, Isola P, Efros AA. Unpaired image-to-image translation using cycle-consistent adversarial networks. arXiv. arXiv preprint arXiv: 1703.10593. 2017. DOI: 10.48550/arXiv.1703.10593

[76] Schlegl T, Seebock P, Waldstein SM, Schmidt-Erfurth U, Langs G. Unsupervised anomaly detection with generative adversarial networks to guide marker discovery. In: Information Processing in Medical Imaging, Lecture Notes in Computer Science. Cham: Springer; 2017. pp. 146-147. DOI: 10.1007/978-3-319-59 050-9.12

[77] Schlegl T, Seebock P, Waldstein SM, Langs G, Schmidt-Erfurth U. f-AnoGAN: Fast unsupervised anomaly detection with generative adversarial networks. Medical Image Analysis. 2019;**54**:30-44. DOI: 10.1016/j.media.2019.01.010

[78] Zenati H, Foo CS, Lecouat B, Manek G, Chandrasekhar VR. Efficient GAN-based anomaly detection. arXiv. arXiv preprint arXiv:1902.03984. 2018, 2019. DOI: 10.48550/arXiv.1902.03984

[79] Akcay S, Atapour-Abarghouei A, Breckon TP. GANomaly: Semi-supervised anomaly detection via adversarial training. In: Lecture Notes in Computer Science. Berlin, Germany: Springer; 2019. pp. 622-637. DOI: 10.1007/978-3-030-20893-6 39

[80] Bashar MA, Nayak R. TANoGAN: Time series anomaly detection with generative adversarial networks. In: 2020 IEEE Symposium Series on Computational Intelligence (SSCI). Piscataway, NJ, USA: IEEE Publishing; 2020. pp. 1778-1785. DOI: 10.1109/SSCI47803.2020.9308512

[81] Li D, Chen D, Jin B, Shi L, Goh J, Ng SK. MADGAN: Multivariate anomaly detection for time series data with generative adversarial networks. In: Lecture Notes in Computer Science. Berlin, Germany: Springer; 2019. pp. 703-716. DOI: 10.1007/978-3-030-30490-4 56

[82] Geiger A, Liu D, Alnegheimish S, Cuesta-Infante A, Veeramachaneni K. TadGAN: Time series anomaly detection using generative adversarial networks. In: Proceedings - 2020 IEEE International Conference on Big Data. Piscataway, NJ, USA: IEEE Publishing; 2020. pp. 33-43. DOI: 10.1109/bigdata50022.2020.9378139

[83] Sorin V, Barash Y, Konen E, Klang E. Creating artificial images for radiology applications using generative adversarial networks (GANs) – A systematic review. Acta Radiologica. 2020;**27**:1175-1185. DOI: 10.1016/j.acra.2019.12.024

[84] Hosny A, Parmar C, Quackenbush J, Schwartz LH, Aerts HJWL. Artificial intelligence in radiology. Nature

Reviews. Cancer. 2018;**18**:500-510. DOI: 10.1038/s41568-018-0016-5

[85] Festag S, Denzler J, Spreckelsen C. Generative adversarial networks for biomedical time series forecasting and imputation. Journal of Biomedical Informatics. 2022;**129**:104058. DOI: 10.1016/j.jbi.2022.104058

[86] Islam J, Zhang Y. GAN-based synthetic brain PET image generation. Brain Informatics. 2020;**7**(1):3. DOI: 10.1186/s40708-020-00104-2

[87] Hirte AU, Platscher M, Joyce T, Heit JJ, Tranvinh E, Federau C. Realistic generation of diffusion-weighted magnetic resonance brain images with deep generative models. Magnetic Resonance Imaging. 2021;**81**:60-66. DOI: 10.1016/j.mri.2021.06.001

[88] Thambawita V et al. DeepFake electrocardiograms using generative adversarial networks are the beginning of the end for privacy issues in medicine. Scientific Reports. 2021;**11**:21896. DOI: 10.1038/s41598-021-01295-2

[89] Lashgari E, Liang D, Maoz U. Data augmentation for deep-learning-based electroencephalography. Journal of Neuroscience Methods. 2020;**346**: 108885. DOI: 10.1016/j.jneumeth.2020.108885

[90] Cheng D, Qiu N, Zhao F, Mao Y, Li C. Research on the modality transfer method of brain imaging based on generative adversarial network. Frontiers in Neuroscience. 2021;**15**: 655019. DOI: 10.3389/fnins.2021.655019

[91] Yurt M, Dar SU, Erdem A, Erdem E, Oguz KK, Çukur T. mustGAN: Multi-stream generative adversarial networks for MR image synthesis. Medical Image Analysis. 2021;**70**:101944. DOI: 10.1016/j.media.2020.101944

[92] Jin CB, Kim H, Liu M, Jung W, Joo S, Park E, et al. Deep CT to MR synthesis using paired and unpaired data. Sensors. 2019;**22**(19):2361. DOI: 10.3390/s19102361

[93] Lan H, Alzheimer Disease Neuroimaging Intitiative, Toga AW, Sepehrband F. Three-dimensional self-attention conditional GAN with spectral normalization for multimodal neuroimaging synthesis. Magnetic Resonance in Medicine. 2021;**86**: 1718-1733. DOI: 10.1002/mrm.28819

[94] Zhao K, Zhou L, Gao S, Wang X, Wang Y, Zhao X, et al. Study of low-dose PET image recovery using supervised learning with CycleGAN. PLoS One. 2020;**15**:e0238455. DOI: 10.1371/journal.pone.0238455

[95] Sundar LKS, Iommi D, Muzik O, Chalampalakis Z, Klebermass EV, Hienert M, et al. Conditional generative adversarial networks aided motion correction of dynamic 18F-FDG PET brain studies. Journal of Nuclear Medicine. 2021;**62**:871-880. DOI: 10.2967/jnumed.120.248856

[96] Delannoy Q, Pham CH, Cazorla C, Tor-Díez C, Dollé G, Meunier H, et al. SegSRGAN: Super-resolution and segmentation using generative adversarial networks—Application to neonatal brain MRI. Computers in Biology and Medicine. 2020;**120**:103755. DOI: 10.1016/j.compbiomed.2020.103755

[97] Shaul R, David I, Shitrit O, Raviv TR. Subsampled brain MRI reconstruction by generative adversarial neural networks. Medical Image Analysis. 2020;**65**:101747. DOI: 10.1016/j.media.2020.101747

[98] An Y, Lam HK, Ling SH. Auto-denoising for EEG signals using generative adversarial network. Sensors. 2022;**22**:1750. DOI: 10.3390/s22051750

[99] Wolterink JM, Leiner T, Viergever MA, Isgum I. Generative adversarial networks for noise reduction in low-dose CT. IEEE Transactions on Medical Imaging. 2017;**36**(12):2536-2545. DOI: 10.1109/TMI.2017.2708987

[100] Sille R, Choudhury T, Sharma A, Chauhan P, Tomar R, Sharma D. A novel generative adversarial network-based approach for automated brain tumour segmentation. Medicina. 2023;**59**(1):119. DOI: 10.3390/medicina59010119

[101] Yuan W, Wei J, Wang J, Ma Q, Tasdizen T. Unified generative adversarial networks for multimodal segmentation from unpaired 3D medical images. Medical Image Analysis. 2020; **64**:101731. DOI: 10.1016/j.media.2020.101731

[102] Oh KT, Lee S, Lee H, Yun M, Yoo SK. Semantic segmentation of white matter in FDG-PET using generative adversarial network. Journal of Digital Imaging. 2020;**33**:816-825. DOI: 10.1007/s10278-020-00321-5

[103] Elazab A, Wang C, Gardezi SJS, Bai H, Hu Q, Wang T, et al. GP-GAN: Brain tumor growth prediction using stacked 3D generative adversarial networks from longitudinal MR images. Neural Networks. 2020;**132**:321-332. DOI: 10.1016/j.neunet.2020.09.004

[104] Han C, Rundo L, Murao K, Noguchi T, Shimahara Y, Milacski ZÁ, et al. MADGAN: Unsupervised medical anomaly detection GAN using multiple adjacent brain MRI slice reconstruction. BMC Bioinformatics. 2021;**22**(Suppl 2):31. DOI: 10.1186/s12859-020-03936-1

[105] Ren Z, Li J, Xue X, Li X, Yang F, Jiao Z, et al. Reconstructing seen image from brain activity by visually-guided cognitive representation and adversarial learning. NeuroImage. 2021; **228**:117602. DOI: 10.1016/j.neuroimage.2020.117602

[106] Huang W, Yan H, Wang C, Yang X, Li J, Zuo Z, et al. Deep natural image reconstruction from human brain activity based on conditional progressively growing generative adversarial networks. Neuroscience Bulletin. 2021;**37**:369-379. DOI: 10.1007/s12264-020-00613-4

[107] Al-Tahan H, Mohsenzadeh Y. Reconstructing feedback representations in the ventral visual pathway with a generative adversarial autoencoder. PLoS Computational Biology. 2021;**17**:1-19. DOI: 10.1371/journal.pcbi.1008775

[108] Goldberger AL, Goldberger ZD, Shvilkin A. Golberger's Clinial Electrocardiography: A Simplified Approach. 9th ed. Philadelphia, PA, USA: Elsevier; 2017. DOI: 10.1016/C2014-0-03319-9

[109] Skandarani Y, Lalande A, Afilalo J, Jodoin PM. Generative adversarial networks in cardiology. The Canadian Journal of Cardiology. 2022;**38**:196-203. DOI: 10.1016/j.cjca.2021.11.003

[110] Shin DH, Park RC, Chung K. Decision boundary-based anomaly detection model using improved AnoGAN from ECG data. IEEE Access. 2020;**8**:108664-108674. DOI: 10.1109/ACCESS.2020.3000638

[111] Li W, Tang YM, Yu KM, To S. SLC-GAN: An automated myocardial infarction detection model based on generative adversarial networks and convolutional neural networks with single-lead electrocardiogram synthesis.

Information Sciences. 2022;**589**:738-750. DOI: 10.1016/j.ins.2021.12.083

[112] Xia Y, Xu Y, Chen P, Zhang J, Zhang Y. Generative adversarial network with transformer generator for boosting ECG classification. Biomedical Signal Processing and Control. 2023;**80**: 104276. DOI: 10.1016/j.bspc.2022.104276

[113] Rath A, Mishra D, Panda G, Satapathy SC. Heart disease detection using deep learning methods from imbalanced ECG samples. Biomedical Signal Processing and Control. 2021;**68**: 102820. DOI: 10.1016/j.bspc.2021.102820

[114] Qin J, Gao F, Wang Z, Wong DC, Zhao Z, Relton SD, et al. A novel temporal generative adversarial network for electrocardiography anomaly detection. Artificial Intelligence in Medicine. 2023;**136**:102489. DOI: 10.1016/j.artmed.2023.102489

[115] Wang Z, Stavrakis S, Yao B. Hierarchical deep learning with generative adversarial network for automatic cardiac diagnosis from ECG signals. Computers in Biology and Medicine. 2023;**155**:106641. DOI: 10.1016/j.compbiomed.2023.106641

[116] Nunez PL, Srinivasan R. Electric Fields of the Brain: The Neurophysics of EEG. 2nd ed. Oxford, UK: Oxford University Press; 2006. DOI: 10.1093/acprof:oso/9780195050387.001.0001

[117] Habashi AG, Azab AM, Eldawlatly S, Aly GM. Generative adversarial networks in EEG analysis: An overview. Journal of Neuro-engineering and Rehabilitation. 2023;**20**:40. DOI: 10.1186/s12984-023-01169-w

[118] Wei Z, Zou J, Zhang J, Xu J. Automatic epileptic EEG detection using convolutional neural network with improvements in time-domain. Biomedical Signal Processing and Control. 2019;**53**:101551. DOI: 10.1016/j.bspc. 2019.04.028

[119] You S, Cho BH, Yook S, Kim JY, Shon YM, Seo DW, et al. Unsupervised automatic seizure detection for focal-onset seizures recorded with behind-the-ear EEG using an anomaly-detecting generative adversarial network. Computer Methods and Programs in Biomedicine. 2020;**193**:105472. DOI: 10.1016/j.cmpb.2020.105472

[120] Tomson T, Nashef L, Ryvlin P. Sudden unexpected death in epilepsy: Current knowledge and future directions. Lancet Neurology. 2008;**7**: 1021-1031. DOI: 10.1016/S1474-4422(08)70202-3

[121] Usman SM, Khalid S, Bashir Z. Epileptic seizure prediction using scalp electroencephalogram signals. Biocybernetics and Biomedical Engineering. 2021;**41**:211-220. DOI: 10.1016/j.bbe.2021.01.001

[122] Natu M, Bachute M, Gite S, Kotecha K, Vidyarthi A. Review on epileptic seizure prediction: Machine learning and deep learning approaches. Computational and Mathematical Methods in Medicine. 2022;**2022**: 7751263. DOI: 10.1155/2022/7751263

[123] Zhu B, Shoaran M. Unsupervised domain adaptation for cross-subject few-shot neurological symptom detection. In: International IEEE/EMBS Conference on Neural Engineering. Piscataway, NJ, USA: IEEE Publishing; 2021. DOI: 10.1109/NER49283.2021.9441235

[124] Truong ND, Kuhlmann L, Bonyadi MR, Querlioz D, Zhou L, Kavehei O. Epileptic seizure forecasting

with generative adversarial networks. IEEE Access. 2019;**7**:143999-144009. DOI: 10.1109/ACCESS.2019.2944691

[125] Pascual D, Amirshahi A, Aminifar A, Atienza D, Ryvlin P, Wattenhofer R. EpilepsyGAN: Synthetic epileptic brain activities with privacy preservation. IEEE Transactions on Biomedical Engineering. 2021;**68**(8): 2435-2446. DOI: 10.1109/TBME.2020.3042574

[126] Yin X, Han Y, Sun H, Xu Z, Yu H, Duan X. Multi-attention generative adversarial network for multivariate time series prediction. IEEE Access. 2021;**9**:57351-57363

[127] Usman SM, Khalid S, Bashir S. A deep learning based ensemble learning method for epileptic seizure prediction. Computers in Biology and Medicine. 2021;**136**:1104710. DOI: 10.1016/j.compbiomed.2021.104710

[128] Salazar A, Vergara L, Safont G. Generative adversarial networks and Markov random fields for oversampling very small training sets. Expert Systems with Applications. 2021; **163**:113819. DOI: 10.1016/j.eswa.2020.113819

[129] Yin X, Han Y, Xu Z, Liu J. VAECGAN: A generating framework for longterm prediction in multivariate time series. Cybersecurity. 2021;**4**:22. DOI: 10.1186/s42400-021-00090-w

[130] Rasheed K, Qadir J, O'Brien TJ, Kuhlmann L, Razi A. A generative model to synthesize EEG data for epileptic seizure prediction. IEEE Transactions on Neural Systems and Rehabilitation Engineering. 2020;**29**:2322-2332. DOI: 10.1109/TNSRE.2021.3125023

[131] Gang D, Alkhachroum A, Bicchi MAM, Jagged JR, Cajigas I, Chen ZS. Deep learning for robust detection of interictal epileptiform discharges. Journal of Neural Engineering. 2021;**18**:056015. DOI: 10.1088/1741-2552/abf28e

[132] Luo TJ, Fan Y, Chen L, Guo G, Zhou C. EEG signal reconstruction using a generative adversarial network with Wasserstein distance and temporal-spatial-frequency loss. Frontiers in Neuroinformatics. 2020;**14**:15. DOI: 10.3389/fninf.2020.00015

[133] Wang J, Mu W, Wang A, Wang L, Han J, Wang P, et al. Generative adversarial networks for electroencephalogram signal analysis: A mini review. In: International Winter Conference on Brain Computer Interface (BCI). Piscataway, NJ, USA: IEEE Publishing; 2023. DOI: 10.1109/BCI57258.2023.10078666

[134] Handa P, Gupta E, Muskan S, Goel N. A review on software and hardware developments in automatic epilepsy diagnosis using EEG datasets. Expert Systems. 2023:e13374. DOI: 10.1111/exsy.13374

[135] Daoud H, Bayoumi M. Generative adversarial network based semi-supervised learning for epileptic focus localization. In: IEEE International Conference on Bioinformatics and Biomedicine (BIBM). Piscataway, NJ, USA: IEEE Publishing; 2021. DOI: 10.1109/BIBM52615.2021.9669695

[136] Dong Z, Zhou S. EEG-based seizure detection using generative model and deep learning. In: IEEE International Conference on E-Health and Bioengineering (EHB). Piscataway, NJ, USA: IEEE Publishing; 2022. DOI: 10.1109/EHB55594.2022.9991438

[137] Ganti B, Chaitanya G, Balamurugan S, Nagaraj N,

Balasubramanian K, Pati S. Time-series generative adversarial network approach of deep learning improves seizure detection from the human thalamic SEEG. Frontiers in Neurology. 2022;**13**: 755094. DOI: 10.3389/fneur.2022.755094

[138] Xu M, Jie J, Zhou W, Zhou H, Jin S. Synthetic epileptic brain activities with TripleGAN. Computational and Mathematical Methods in Medicine. 2022;**2022**:2841228. DOI: 10.1155/2022/2841228

[139] Zhang X, Yao L, Dong M, Liu Z, Zhang Y, Li Y. Adversarial representation learning for robust patient-independent epileptic seizure detection. IEEE Journal of Biomedical and Health Informatics. 2020;**24**(10): 2852-2859. DOI: 10.1109/JBHI.2020.2971610

[140] Boonyakitanont P, Lek-uthai A, Chomtho K, Songsiri J. A review of feature extraction and performance evaluation in epileptic seizure detection using EEG. Biomedical Signal Processing and Control. 2020;**57**:101702. DOI: 10.1016/j.bspc.2019.101702

[141] Cherian R, Kanaga EG. Theoretical and methodological analysis of EEG based seizure detection and prediction: An exhaustive review. Journal of Neuroscience Methods. 2022;**369**: 109483. DOI: 10.1016/j.jneumeth.2022.109483

[142] Nafea MS, Ismail ZH. Supervised machine learning and deep learning techniques for epileptic seizure recognition using EEG signals—A systematic literature review. Bioengineering. 2022;**9**:781. DOI: 10.3390/bioengineering9120781

[143] Yuan J, Ran X, Liu K, Yao C, Yao Y, Wu H, et al. Machine learning applications on neuroimaging for diagnosis and prognosis of epilepsy: A review. Journal of Neuroscience Methods. 2022;**368**:109441. DOI: 10.1016/j.jneumeth.2021.109441

[144] Xu Q, Huang G, Yuan Y, Guo C, Sun Y, Wu F, et al. An empirical study on evaluation metrics of generative adversarial networks. arXiv. arXiv preprint arXiv:1806.07755. 2018. DOI: 10.48550/arXiv.1806.07755

[145] Borji A. Pros and cons of GAN evaluation measures: New developments. Computer Vision and Image Understanding. 2022;**215**:103329. DOI: 10.1016/j.cviu.2021.103329

Chapter 3

Anomaly Detection in IoT: Recent Advances, AI and ML Perspectives and Applications

Menachem Domb, Sujata Joshi and Arulmozhi Khn

Abstract

IoT comprises sensors and other small devices interconnected locally and via the Internet. Typical IoT devices collect data from the environment through sensors, analyze it and act back on the physical world through actuators. We can find them integrated into home appliances, Healthcare, Control systems, and wearables. This chapter presents a variety of applications where IoT devices are used for anomaly detection and correction. We review recent advancements in Machine/Deep Learning Models and Techniques for Anomaly Detection in IoT networks. We describe significant in-depth applications in various domains, Anomaly Detection for IoT Time-Series Data, Cybersecurity, Healthcare, Smart city, and more. The number of connected devices is increasing daily; by 2025, there will be approximately 85 billion IoT devices, spreading everywhere in Manufacturing (40%), Medical (30%), Retail, and Security (20%). This significant shift toward the Internet of Things (IoT) has created opportunities for future IoT applications. The chapter examines the security issues of IoT standards, protocols, and practical operations and identifies the hazards associated with the existing IoT model. It analyzes new security protocols and solutions to moderate these challenges. This chapter's outcome can benefit the research community by encapsulating the Information related to IoT and proposing innovative solutions.

Keywords: anomaly detection, internet of things [IoT], cybersecurity, data security, threats, risks, smart devices, time-series data, AI, machine learning, deep learning, healthcare, smart city, IoT environments, internet of things, anomaly detection, IoT intrusion detection, machine learning, deep learning, transfer learning, network security, convolutional neural network

1. Introduction

The wide variety of IoT devices lacking any standard creates connectivity issues and increases the security vulnerability of IoT local networks and the entire Internet. Machine Learning techniques are already used in ECG, X-ray, pattern recognition, cancer detection, brain signal modeling, and IoT services on electrical impedance planes to discover defects. Extending ML and DL technologies to detect anomalies where it is already operating is a natural and effective transition. Anomalies are events or patterns that deviate significantly from predictable behavior. Detection methods are expected to identify anomaly occurrences and their probable cause promptly. To comply with this chapter topic, we focus on these applications incorporating Machine Learning and Deep learning methods. Chatterjee & Ahmed [1] provide a comprehensive survey on Anomaly Detection in IoT and propose four measurements for evaluating IoT Anomaly Detection methods: how they approach the problem, how they are applied, the type of method, and the algorithm latency. Anomaly detection using deep learning is described by Chalapathy and Chawla [2], and Yassine et al. [3] provide a review of the methodologies, situations, and computation platforms used for anomaly detection in the energy industry. Talagala et al. [4] propose a distributional unsupervised for anomaly detection in high-dimensional data. Yin et al. [5] extract unique temporal features from a given temporal data file using a combination of CNN and LSTM and continue in [5] to detect anomalies involving CNN, LSTM, and Deep neural network (DNN).

They [1] also define 18 application types of anomaly detection processes. The following are examples of various application types. Sobhani et al. [6] demonstrate that the accuracy of final load projections is improved when eliminating observations from the original input using local load information. T. Asakura et al. [7] detect damage to industrial rotating equipment by calculating the feature vectors of the anomaly vibration data extracted from sensors' vibration signal features to construct a monitoring system for machinery equipment. Huang et al. [8] proposed anomalies detection in Manufacturing using density peak weighted fuzzy C-means (WFCM). Yasaei et al. [9] detect unexpected event changes in sensor signals using an adaptive data-driven monitoring method. Zekry et al. [10] use a convolutional LSTM model for anomaly detection in the context of connected vehicles. Wang et al. [11] log anomalies in IoT systems using a natural language processing approach, extracting the relevance between words and vectoring them. The method trains supervised models to detect anomalies reducing computational time. Xu et al. [12] used I-LSTM and Deep learning in smart-city data for multi-classification anomaly detection to improve smart homes' service quality. Tripathi et al. [13] proposed reliable and transparent city connectivity using IoT, MEC, and Blockchain consensus. Ullah et al. [14] presented a timely identification of abnormal incidents in surveillance networks, incorporating LSTM with CNN, where CNN features are collected from successive frames. LSTM is used to distinguish between normal and abnormal values. The in-depth features and multi-layer BD-LSTM provide high-level training and validation data to real-world IoT surveillance networks. The DeL-IoT framework [15] detects IoT abnormalities by observing flow-level traffic instances that pass through switches. The IoT anomaly identification and prediction framework uses a Deep Learning technique to identify anomalies. Mirsky et al. [16] proposed a Blockchain-based distributed anomaly detection algorithm using the Markov chain (MC) to simulate sequences efficiently. Y. An et al. [17] proposed anomaly detection capable of relieving network congestion and CPUs from the computing pressures of centralized servers, unlocking the potential of edge intelligence in IoT. Shen et al. [18] propose a privacy-preserving SVM training strategy using encrypted IoT data. Data providers encrypt their data locally using their private keys and then record the encrypted data on the Blockchain.

The rest of the chapter comprises as follows: The next section outlines security issues unique to the IoT environment. Section 3 presents a generic two-stage Anomaly Detection approach. In the first stage, a process builds the envelope around the weighted average, and the comparison is done in the second stage. In Section 4,

Anomaly Detection using Random Forest Machine Learning is presented, and it concludes in Section 5.

2. IoT security issues

We see a considerable rise in the use of IoT applications in our day-to-day lives. The IoT enhances web-based applications by enabling connections via the Internet between people and their equipment/devices in a real-world or virtual environment. IoT improves Web-enabled applications by allowing links between "everyone" and "everything" in a real-world and virtual environment [19]. Utilizing IoT applications and services is now easier than ever because of the exponential expansion of smart devices. As the asset value of the data kept, processed, and conveyed increases along with scale, so do the attacks against them. These predictions show that there will be a rise in the number and level of threats and attacks against IoT devices, necessitating more robust security measures. This section aims to investigate recent IoT cybersecurity solutions.

Artificial intelligence, Machine/Deep Learning, and networking have become the current area of IoT-related research. Adopting ultra-lightweight protocols for security and core functionality is a significant development in the IoT.

IoT security is constantly evolving, with new risks always being found. The focus of IoT security discussions is ACL techniques, interim encryption techniques, hardware-specific security solutions, and SQL-related attack measures. Identifying IoT-related cybersecurity risks, providing classifications, and looking for prevailing solutions to address them. The following questions are addressed:

1. What architecture can be used considering various criteria?

2. What are the IoT standards and protocols currently in use?

3. What are the IoT cyber security threats?

2.1 Literature review

The recent industrial trends include embedded networking in the wireless segment, where IoT is the major player. The demand for smart applications and systems grew, leading to the rise of IoT in commercial segments [20]. Due to the immense increase in the retail segments, the usage of smart applications has spiked up, increasing their dependability, which further leads to high risks. IoT devices have emerged as the spot for intrusion activities because of the lightweight protocols and standards that are currently present on these devices [21, 22], and the entities that make up these devices have easier access to servers [23] because the security is not fully resolved. The problem with the traditional model is the lack of low-powered device algorithms and the incompatibility of security tools due to differences in policy and implementation methods [24]. A variety of hardware-based techniques and unique solutions have been suggested in recent research to address traditional security challenges.

Xin Zhang and Fengtong Wen [25] proposed an authentication for IoT, where two algorithmic models have been built to ensure valid authentication. The scope of the security solution offered in this work is constrained to protect only lightweight sensor devices from the standard network layer and physical layer-based attacks. M. Dahman Alshehri and Farookh K. Hussain [26] proposed a cluster-based fuzzy architecture and a secured communications model for IoT nodes. This study effectively provides a detection technique against the network's malicious nodes but does not cater to the threats posed by the audit attack surface. This study does not adequately address the performance analysis of operational communication and computing costs. Chen et al. [27] offered a unique Low scale Denial of-Service attack detection approach that incorporates trust evaluation with Hilbert-Huang Transformation in Zigbee WSN to address the security risks considering a large number of devices with low energy which is susceptible to attacks. This work's signal and anomaly detection technique helps reduce the attack level. It has an extensible design because it supports cloud and edge computing, but higher storage overheads persist as a problem. In traditional network security, IDS is entrusted with identifying and keeping track of threat behaviors. Hence, such models do not expressly target the IoT environment.

2.2 Security architecture and communication

This section discusses the IoT security architecture. Use-cases for IoT range from single node devices to cross platform deployments of technology and real-time cloud systems [28]. IoT operations consist of three main tasks: transmitting, retrieving, and data processing. Application Layer: Embedded interface modules enable devices to communicate with the underlying architecture. The device Management Plane identifies the data's source and destination to maintain the device's input–output operations. For instance, the Aggregator aggregates the given device data assets into a fixed set. A communication Layer is an intermediary layer with network components that establish various communication protocols and standards. This layer comprises stacks of current protocols and criteria for controlling traffic throughout the system. Standard protocols enable proper communication among IoT devices. Such systems need a defined set of simple rules to initialize and share data information. **Figure 1** depicts the multi-layer architecture of IoT.

The IoT's communication protocols include:

1. Z-wave – The protocol facilitates device communication in a closed network. It implies that the Z-wave regulating code is not publicly accessible. It prohibits anyone from changing the code and suggests that each Z-wave device has a unique ID granting access to all remote controls. This architecture ensures effective interoperability and security, the Z-wave protocol's core.

2. BLE –Bluetooth low energy is a widely used protocol. Due to its propensity to consume less energy, it works well with low-energy devices. Based on Generic Attributes, this protocol uses Services and Characteristics to carry out its operations.

3. MQ Telemetry Transport (MQTT) is a protocol for small Internet of Things (IoT) devices that allow data transmission and some reception between the sensor nodes.

4. Advanced Message Queueing Protocol (AMQP): includes efficiency, portability, multichannel support, and security, a TCP-dependent binary protocol that ensures authentication using SASL or TLS.

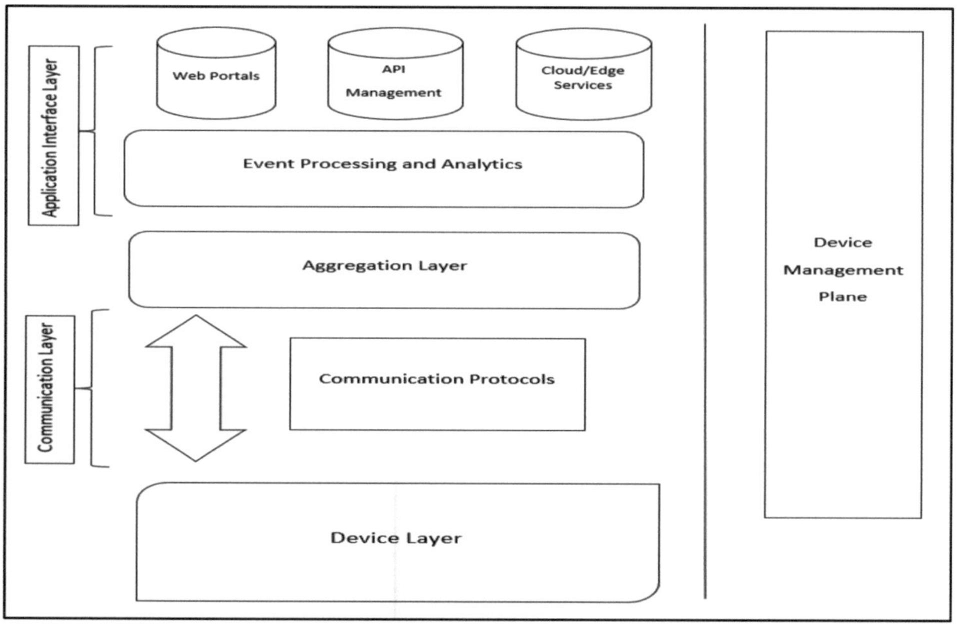

Figure 1.
IoT multi-layer architecture.

5. Limited Application Protocol (CoAP) is a protocol for constrained-based environments. Significant traits of this protocol include its REST API-based foundation, design for system applications, effective congestion control, cross-protocol interoperability, and many others.

6. The Data Distribution Service protocol is an Internet of Things protocol for M2M communications. Like the MQTT and CoAP protocols, data can be exchanged using the publish-subscribe approach; the significant distinction is that this architecture does not require a broker, unlike the latter two. DDS employs multicasting to provide apps with high QoS.

7. 6LoWPAN is the 6th version of the Low-power Wireless Personal Area Network. It is a standard protocol for implementing IPv6 on wireless networks comprising low-power wireless modules.

8. DTLS: Datagram Transport Layer Security is a security protocol for the Internet of Things and is intended to safeguard data transmission between apps that use datagrams. It offers the same level of security and is majorly focused on the Transport layer security protocol.

Heterogeneous physical components such as switches, actuators, gateways, sensor nodes, and other embedded devices make up this unstable environment. A significant impact on networking principles is made by the intelligent device's engineering process, which is the backbone of the whole concept. Gadgets with self-configuring capabilities of the M2M communication paradigm are IoT innovations. Through algorithms and auxiliary technology, this configuration gives nodes the intelligence they need to make decisions for themselves under any circumstance [29, 30]. It is helpful during rescue operations and other emergencies where configuring the network for a specific area is complex, and there is no support for damaged nodes. However, as machines are not failsafe, it becomes susceptible if it depends too heavily on them. Particularly in the present, adversaries use weak authentication, unpatched firmware, and online credential vulnerabilities [31].

Following are some of the IoT security issues:

1. Heterogeneous devices: the paradigm most sensitive to access requests, detecting third-party indulgence, and limited scalability compliance with security management. Several security issues with IoT today relate to traditional network architecture, including IoT devices that interact with the physical environment differently than conventional network devices did in the past. IoT devices' heterogeneous nature ramifies other components as they operate. NIST stressed that IoT-specific privacy regulations [32] and cyber controls must consider the consequences that impact physical systems [33], ultimately affecting the physical world.

2. Regulations & Policies: No global IoT security standard applies to all IoT industry segments. Although some regulations are in the process (such as the EU's General Data Protection Regulation and the US IoT Cybersecurity Improvement Act), they are relatively fragmented and do not address issues with IoT. IoT devices are used globally by many servers, whether they are in a business/ in a person's workspace. Such devices can be monitored/managed using a different rule engine, and the security policy varies depending on the system's devices. Therefore, regularization requires updating every device, which is time-consuming and challenging for any company. Problems include an un-uniform pace of updating, new switches leaving some devices behind that are not updated, or inadequately configured nodes since it takes time to maintain track of millions of nodes.

3. Additional Plugins and Security: Since providing security measures for IoT was never modeled, further security controls are added to the IoT's security architecture. Unlike the traditional networking paradigm, the effectiveness of security characteristics relies on the IoT architecture's ability to function with additional resources. The efficiency of the IoT's security is also influenced by client behaviors, such as how they choose among the various security solutions.

4. Lack of compliance: The lack of compliance among manufacturers is always a cause for concern. A device should generally satisfy the following requirements: Operational Compliance, Security Compliance, and Manufacturing Compliance.

5. An IoT network may be in danger if operational compliance is not maintained. A city's power distribution login system could be part of the network. The network on which they operate is at risk due to legacy operating systems which delayed security patches and other issues. Few makers of IoT devices utilize open-source code. When these IoT devices join a network, the entire system's integrity may be compromised. A lack of security compliance only makes IoT security issues more difficult. Many IoT device producers need to create patchable IoT products.

6. Exposure Threats: IoT endpoints, such as sensors and IP cameras which are in public spaces, are the threat points that are easiest for the enemy to access. As a result, the user's integrity and authentication are threatened by physical-based and proximity threats [34]. Our changes to the protocol method to safeguard devices from adversaries are the biggest security difficulties in this area.

2.3 Classification of IoT attacks

Several commercial businesses have made significant financial investments to secure their IoT-based networks in recent years. IoT attacks are split into two modules:

2.3.1 Protocol-based attacks

Protocol-based attacks utilize known published protocols to serve their benefits, affecting the communication channel. It is divided into two types:
1. Communications protocols attacks: (a) Attacks on communication protocols—several types of exploitation occur when nodes transition, such as sniffer attacks, flooding attacks, and key preshredding attacks. (b) Network protocol attacks where connection establishment is exploited include Wormhole attacks, selective forward attacks, and sniffing attacks.

2.3.2 Transmitted data attacks

Threats on initial packets and messages moving across communication nodes. Some of its most severely affected security exploitations are data leakage, malicious node VM formation, hash collision, and denial of service. Active and passive attacks compromise the system's security—the effectiveness of the network is less affected by passive attack protection systems, which are restricted to monitoring techniques. Modern, responsive security techniques are needed to counter active attacks to reduce risk and affect network performance.

a. Distributed Denial of Service attack — DDoS [35] impacts a network security parameter's availability. Botnets enable DDoS threats on sensor nodes. Affected packets from various sources get access via these points, travel down network data routes, and end up clogging the entire link architecture, making servers unusable.

b. Sniffing attack [36] falls under data collection, a threat vector in which vital system data is collected and used for attacks. With the use of sophisticated tools, information assets are examined. Most devices available on the market need to be sufficiently clever to counteract and are mainly targeted by them.

c. Replay Attack – A replay attack consists of the following steps: "eavesdropping on the communication link between IoT devices or the gateway; intercepting the acknowledgments or connection-establishing components; and deceitfully delaying or rerouting the message."

d. Masquerade attack [37] – This attack impersonates a valid access identification procedure to get access to target node information. Devices that have poor authorization procedures are highly vulnerable. Such attacks use stolen passwords and user credentials by exploiting program gaps or developing workarounds for the current authentication procedure.

e. Port Scanning - Synchronize requests, target ports, sources, firewalls, packets, open nodes [38], and listening nodes. Synchronize scans are a frequently used technique that creates a partial connection to the target node on the target port by sending a synchronized packet to test the host system's initial response.

2.4 IoT security solutions

In contrast to traditional security, which is tool-centric, the most recent cybersecurity solutions focus on software-oriented techniques [39, 40]. The security characteristics that current systems address are authentication, trust, and integrity. Even in its current state, the Internet of Things (IoT) cannot support powerful devices and is not adaptable enough to keep up. **Table 1** summarizes the IoT protocols, emphasizing their characteristics and security concerns. According to the findings, protocol-based

S.No	Protocol used	Features	Cyber Security issues
1	Z-wave	• Z-Wave is a low-power RF technology that can control up to 230 devices at once and builds a wireless mesh network by delivering signals in the sub-1GHz frequency.	An attacker within Z-Wave radio range could control weak devices, deny service, force devices to fail, deplete batteries, intercept, observe, and replay traffic.
		• Minimal interference, reliable connectivity, high security through encryption, and fewer disconnections will be the main advantages of using this IoT Data Protocol.	
2	BLE	• Offers a similar range to traditional Bluetooth.	Susceptible to cyberattacks and interception when sending and receiving data.
		• Has a mesh networking structure.	
		• Designed for low-energy gadgets.	
3	MQTT	• Power usage is comparatively low.	Internet-based MQTT servers that have been exposed, and malicious third-party MQTT message subscriptions.
		• Malicious sinkhole and wormhole attacks against nodes and Distributed Denial of Service (DDoS) assaults.	
		• Provides a simple protocol for TCP data exchange between machines.	
4	AMQP	• Deliveries of messages with reliability, messages delivered quickly, and acknowledgments in messages.	Security of message broker is affected.
		• Most corporate messaging uses AMQP.	

S.No	Protocol used	Features	Cyber Security issues
5	CoAP	• Designed for the limited network device environment.	In a DDoS attack, a third party simultaneously sends forged IP packets during CoAP reflection and amplification.
		• Specialized application for the homogeneous community of restricted devices.	
		• Consists of a variety of end node devices, constrained small networks over the Internet connection.	
6	DDS	• Has a communication protocol that varies from machine to machine.	Because of the expandability feature, poorly implemented and managed devices might result in Man in the Middle or DDoS attacks.
		• High performance	
7	NFC	• Make sure the two-way connection is safe; Usage of smartphones as the end nodes.	Malicious wormhole attack based on nodes.
8	SigFox	• With low-power consumption, it makes the most of both the cellular and WiFi networks.	Poor payload encryption.
		• Supports star network topology and dense node networks.	
		• Has restricted endpoint access control and cloud access.	
9	EnOcean	• Self-powered wireless sensor network that is user-driven and gathers data.	Optional blocking, preshared security keys, and undefinable re-synchronization of rolling codes are frequently overlooked.
		• Key features include less idle current.	
10.	DTLS	• A retransmission timer is used by DTLS to address the packet loss problem. The client retransmits the data if the timer expires before it receives the server's confirmation message.	DDos Attacks.
		• By assigning a unique sequence number to each message, the reordering problem is resolved. This aids in assessing whether or not the subsequent message to be received is in sequence. If it is out of order, it is placed in a queue and dealt with when the appropriate number in the sequence is reached.	
		• DTLS is used in applications where data loss is significantly less essential latency.	

Table 1.
Summary list of security protocols for IoT.

security solutions protect against most IoT attack surfaces [41]. Using secure techniques performed over the Data Link and Transport layers, protocols like COAP and DDS enable efficient immunity against well-known attacks like DDoS attacks and botnet attacks. In Sigfox and EnOcean, new methodologies prevent new threat issues like asynchronous code definition and poor payload encryption. The lightweight protocols MQTT and BLE have also emerged as a viable defense against dangers posed by malicious nodes and Man in Middle attacks. Divided security management is beneficial for more straightforward management of security measures and increases the efficacy of the most suggested solutions.

2.5 Summary

This section discussed IoT's current cyber security trends by researching various protocols, standards, and threats. The research findings on the cyber security risks convey that the traditional methods must be more efficient against attacks in heterogeneous IoT environments. Our study further reveals that most cyber security solutions include encryption techniques with low energy use, which also is successful in securing channel attacks in IoT. IoT security increased after integrating with various technologies.

The complications of the IoT system have increased, and security features' openness has decreased. Even though the previously discussed issues have been attempted to address the evolution of communication technologies and protocols, there is always room for research.

3. Anomaly detection using an optimized envelope

IoT systems collect vast amounts of data to track and analyze the structure of future recorded data. However, this data cannot be stored as is due to limited storage but must be reduced to allow future data analysis based on past data that will not be compromised. We propose a parameterized method of sampling the data optimally. Our approach has three parameters– an averaging process for constructing an average data cycle from past observations, an envelope method for defining an interval around the average data cycle, and an entropy method for comparing new data cycles to the constructed envelope enabling identifying anomalies and predicting future cycle behavior. This section concentrates on finding the optimal envelope using entropy methods.

We often have sequential data collected by sensors, and computational power and bandwidth resources prohibit us from collecting large-scale data. Sampling preserves the most critical information from the original data and reduces the complexity of the subsequent knowledge discovery task to a traceable version without compromising performance. Dictionary learning [42] helps extract patterns hidden in data. We can apply dictionary learning to sequential data for natural language processing, video analysis, and nonsequential data tasks [43]. Given the IoT data collected sequentially, we can find a method that maintains a basis where we have enough elements to describe the sequential patterns of the data. It helps to extract a set of common sequential patterns from the sequential telematics data. In a smart home system, we may collect the most frequent activity trajectories for home members to use for member authentication. We aim to find an optimal sampling method given a set of time-series records, where we collect information before and after the sampling reduction process regarding the data's purpose in the context of the relevant application. Many known data reduction techniques enable restoring the original data set from the reduced one. Among these are compression and compaction routines and dictionary methods. Given the sequential data, we may apply Classification and Prediction. Classification defines whether a series of daily temperatures represent an El-Niño year or whether the data points to suspected intrusion.

3.1 Related work

Vlachos et al. [44] proposed a procedure for getting the best practical estimated gap between two extreme measurements related to any data sequence. Sakurada and Yairi [45] use auto-encoders with nonlinear dimensionality reduction for the anomaly detection task. Reeves et al. [46] generate domain representations using scaleable layers. Chilimbi and Hirzel [47] implement an iterative scheme that uses temporal data to construct a profile. Then, they identify repeated data sequences with the same order, prefetches them, and let the program continue executing the prefetched instructions. Lane and Brodley [48] use instance-based learning (IBL) for boundary determination by good user behavior and heuristics. Kasiviswanathan et al. [49] detect and cluster user content for optimization. Mairal et al. [42] create a dictionary and adapt it to specific data using data vectors proposing an optimization algorithm for dictionary learning based on stochastic approximations. Aldroubi et al. [50] claim that a collection of subspaces gives the best sparse representation providing an optimized sampling in subspaces union. Rubinstein et al. [51] survey the various options up to the most recent contributions and structures. Cherian et al. [52] propose learning over-complete dictionary models where the signal can have both Gaussian and (sparse) Laplacian noise. Dictionary teaching in this setting leads to a complex nonconvex optimization problem, further exacerbated by large input datasets. Duarte-Carvajalino and Sapiro [53] introduce a framework for the joint design and optimization of the nonparametric dictionary and the sensing matrix. They demonstrate the use of random sensing matrices and those optimized independently of the learning of the dictionary. They complement the classical image datasets, maximizing the size of the sampling data to keep the balance between the sampling data and the information extracted from it. Our problem statement focuses on extracting concepts, methods, rules, and measurements so that, at the end of the process, the original sampling data becomes redundant and need no longer be stored. However, we incorporate an ongoing learning process to keep improving and adjusting the extracted artifacts to natural changes in the sampled mechanism's behavior. Our study concentrates on time-dependent streaming sampling data divided by fixed periods to repeat the analysis process for each period/cycle. We propose a condensed and adjustable representation of the data. Reeves et al. [46] offer an alternative to the subject.

3.2 Introducing the envelope approach

Assuming periodic data sampling and extraction of logical artifacts at the period level, we analyze the data collected over several periods. We divide the period into time units. For example, we divide it into daily time units for a year. We average the samples collected during each time unit and extract one value representing it. We repeat this process for the period and get a graph illustrating the average values for an intermediate and typical period. We then calculate the envelope around this average. The generated envelope represents the standard range of values such that unanalyzed periods are compared to this envelope. This period is normal if its graph value is entirely within the envelope. If it is totally out of the envelope, it is an exception.

If just sections of the graph are within the envelope, we use an entropy measure to calculate the "distance" of the given period from the standard envelope. Assuming an existing entropy threshold, we can decide whether the period is typical. We apply the same concept at the unit level and determine whether a specific time unit in a period is within the standard. This particular check is relevant to anomaly detection of IoT behavior. **Figure 2** depicts the main blocks of the envelope construction process.

The process has three key elements: an average measure per time unit, the boundaries around the middle chart, and an entropy value representing the distance of an actual chart from the envelope. We propose an optimal intensity of each component to generate a balanced and reliable anomaly detection method. We start by analyzing typical data collected from several time-dependent cycles, determining the average value per time unit, and drawing the boundaries around the average to get the envelope, as described in detail in **Figure 3**.

Figure 4 describes the anomaly detection process by summing–up the number of cases in the examined chart that exceeds the envelope boundaries and in what direction.

This envelope method is generic and may be used for any application for anomaly detection, such as IoT sensors. In high variations, it can detect damaged or attacked sensors or support automatic instant corrections where abnormal behavior is seen. We may run a backtracking process for ongoing calibration of system parameters. This idea may be used to construct a multi-dimension envelope to comply with dependency among several columns within the same record.

3.3 Experiment

We accepted detailed Meteorological data about El-Niño (EN) and NonEl-Niño years (NEN) from 1980 to 1998. We took data from the El-Niño years 1982, 1983, 1987, 1988, 1991, and 1992 for the positive envelopes. All other years in the range were Non-El-Niño years. We tested three methods for generating envelopes: (1) minimum over all cycles and maximum over all cycles, (2) average cycle ± standard deviation, and (3) confidence interval (CI). **Figure 5** visually confirms that 1995 is a regular year concerning its temperature spread. The Red and Blue charts represent the envelope's

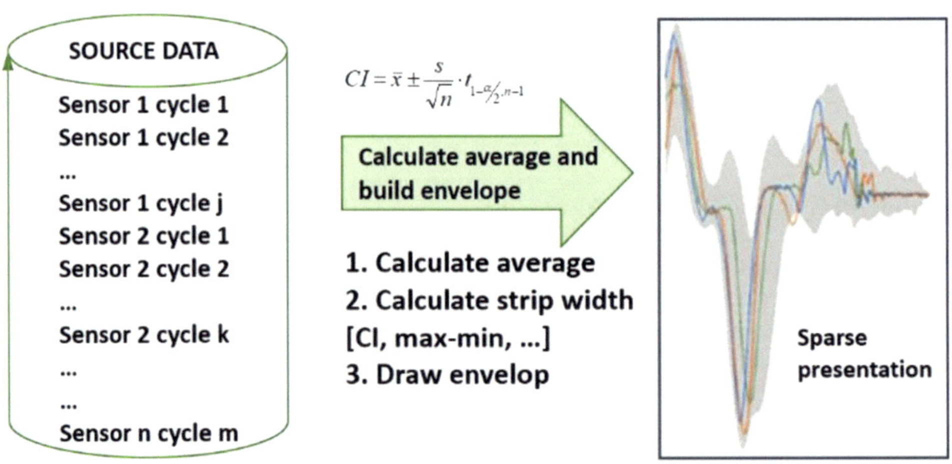

Figure 2.
The process of constructing the optimal envelope.

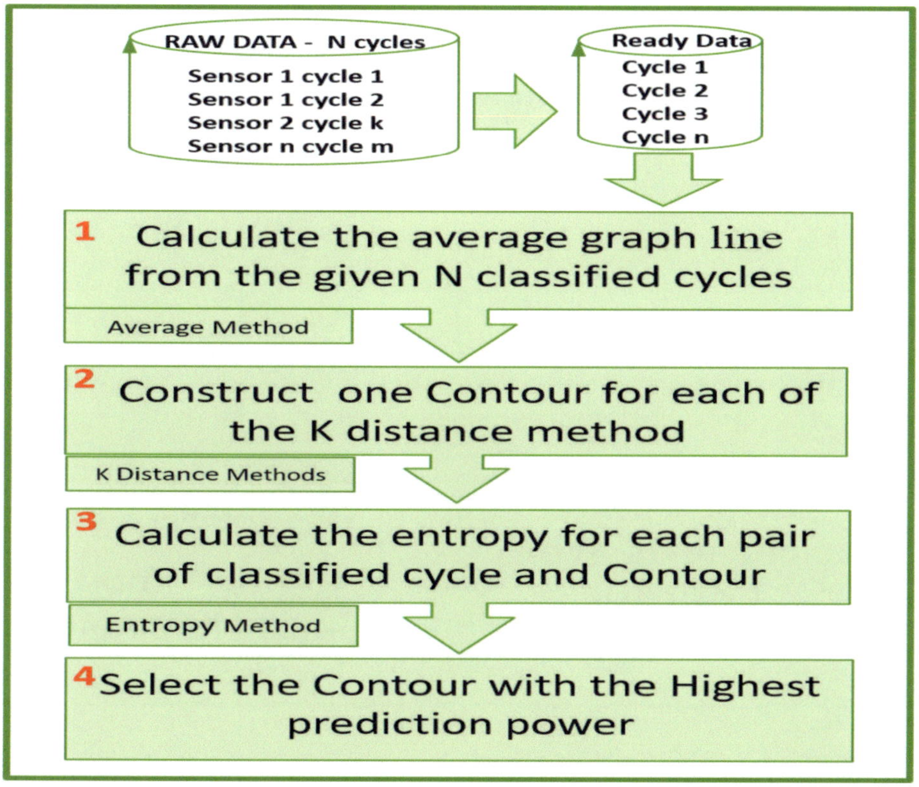

Figure 3.
The process of Constructing the Optimal Envelope.

upper and lower borders, respectively, while the Green chart represents the temperature in 1995. We realize that most temperatures are within the envelope upper/lower boundaries, generating a relatively low Entropy, 0.3631, beneath the selected threshold, concluding that 1995 is indeed a NEN year. However, referring to the 1992 and 1988 years, we got 0.4266 and 0.3857 Entropy values above the threshold; hence they are classified as EN years. However, we did not get a precise classification when we applied the ± standard deviation and the confidence interval (CI) methods.

3.4 Summary

Classification methods have recently gained attention due to rising IoT security issues and threats. In this section, we proposed an envelope construction to classify streams of time-dependent events within a defined data cycle. We discussed three envelope construction options: min–max, standard deviation, and confidence interval (CI). We described an Entropy calculation and a Threshold determination to classify whether a given steam data cycle is abnormal. We used Meteorological data streams to demonstrate our proposal technology's correct classification of daily temperature streams for a year cycle. Several extensions to our proposal include discovering early trends of behavior changes, determining the number of data cycles required for constructing the optimal envelope, exploring the possibility of dividing one cycle into segments associating different envelopes to each segment, and defining rules for anomaly discovery.

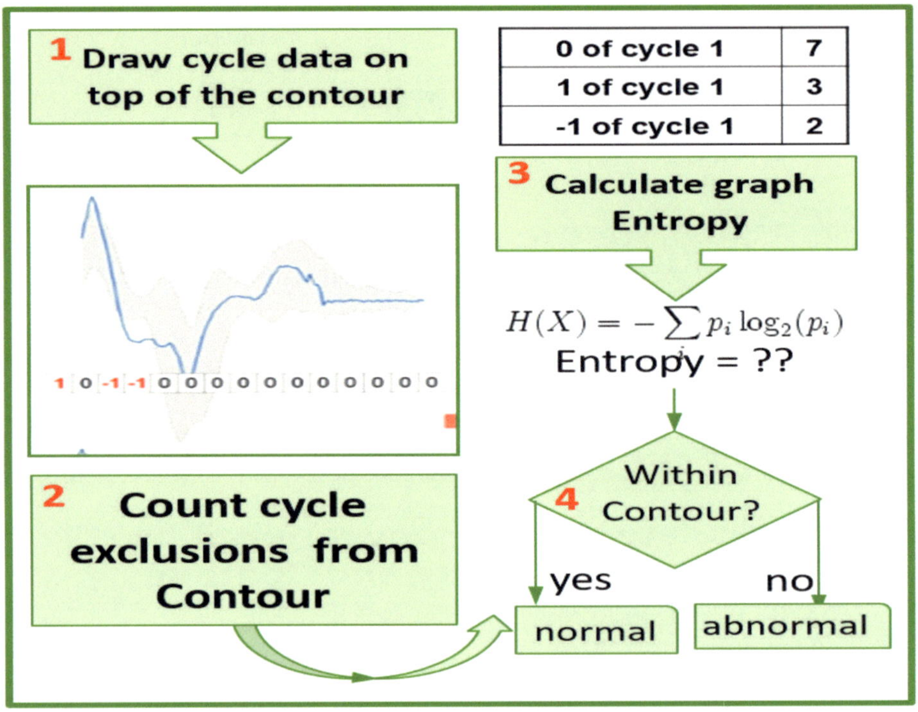

Figure 4.
Classifying an unclassified Cycle.

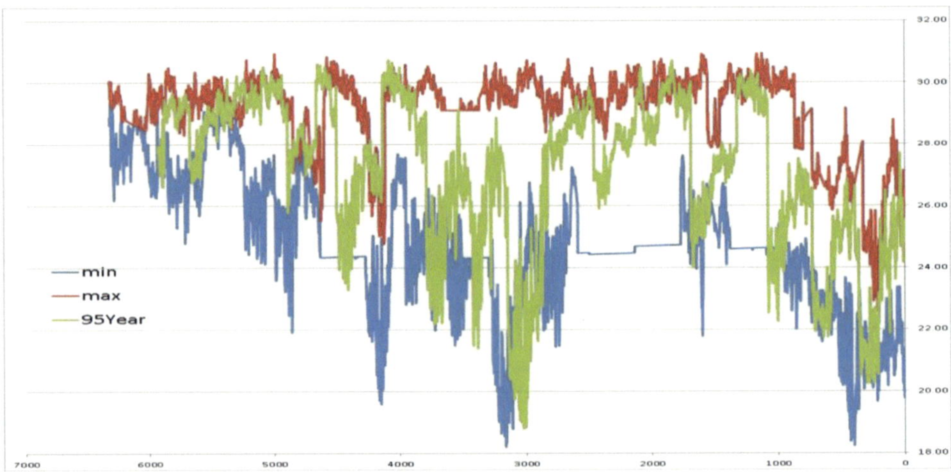

Figure 5.
Min–max envelope for 1995 NEN.

4. Anomaly detection using random forest machine learning

The total transmitted data over the various sensors is growing accordingly. Sensors typically are low in storage, memory, and processing power resources. Data security and privacy are part of this ever-increasing domain's significant concerns and drawbacks. A penetration discovery tool is recommended to predict possible attacks. Machine Training data leads to the definition of good and bad patterns for generating a Lightweight and activation framework comprised of Machine learning rule discovery, threat modeling, and timely reaction to rule violations. The model discovers exceptions and immediately updates the system. Random Forest (RF) is used for anomaly detection and rules generation. We converge IoT groups' resource sharing to build an efficient IoT security framework. IoT networks collect and exchange vast data raising major security issues. To cope with it, we propose a decentralized, layered, distributed, and parallel processing model embedded in the.

The IoT network utilizes the remaining resources to execute the RF method to detect abnormalities. The model supports continued use and is decentralized over time.

The system identifies repeated patterns, while the Machine Learning algorithms discover the geometric, arithmetic, and additive. The patterns are translated into rules to be executed in violation cases. Anan adaptive extension is used to detect changes in generating data and adapt the decision mod to manage suspected situations sel.

The aim is to have a framework with training data collection analyzing it to detect patterns, proportions, etc., and converting it to rules. Combining the collected rules and RF trees is deployed in the IoT devices and network. The rules are executed when data is received from or transmitted to an IoT device. The corresponding action is triggered to cope with the situation if the result is positive or negative.

4.1 Literature review

Eghbal et al. [54] propose analyzing numerical data and generating fuzzy rules. The algorithm uses some rule-and-data-dependent parameters and a function that modifies the rule evaluation measures to assess the candidate rules effectively. Ref. [55] uses Sugeno integrals. They are qualitative criteria aggregations where it is possible to assign weights to criteria groups. It shows how to extract if-then rules expressing the selection of good situations based on local regulations and evaluations to detect bad conditions [56]. Dealing with converting data into the appropriate layout requires a significant investment in manual reformatting. The paper introduces a synthesis engine to extract structured relational data. It uses examples to synthesize a program utilizing an extraction language that extends regular expressions with geometric constructs. Ref. [57] proposes a fast and compact decision rules algorithm. It works online to learn rule sets. It presents a technique to detect local drifts relying on the rule set modularity. Each rule monitors the evolution of performance metrics to detect concept drift. It provides valuable information about the dynamics of the process generating data, faster adaptation to changes, and generates more compact rule sets [58, 59]. It uses averaging techniques to propose a method in which a previous algorithm for association rules mining specifies the minimum support automatically. It uses fuzzy logic to distribute data in different clusters and then tries to introduce to the user the most appropriate threshold automatically [60]. Suggests a two-stage hybrid model for data classification and rule extraction. The first stage uses a Fuzzy ARTMAP classifier with Q-learning and Genetic Algorithm for rule extraction from QFAM. Given a new data sample, the model can provide a prediction about the target class of the data sample and give a fuzzy if-then rule to explain the forecast. Q-values are applied to reduce the number of prototypes generated by QFAM [61]. Proposes a granular-rules extraction method to simplify a data set into a granular-rule set with unique granular rules [62]. It describes a QAR (Quick Access Recorder) anomaly detection algorithm. The method retains the time characteristics data and strengthens the relationship between the condition and decision attributes [63]. Describes an approach of data mining with Excel using the XLMiner add-on. It presents an example of mining association rules to illustrate this approach's steps [64]. Introduces an algorithm for choosing which instances to request next in a setting where the learner can access a pool of unlabelled samples and request some labels [65]. It focuses on understanding the stochastic process's role and how it defines a distribution over functions. It presents the simple equations for incorporating training data and examines how to learn the hyper-parameters using the marginal likelihood [66]. Proposes an active learning algorithm that balances such exploration with refining the decision boundary by dynamically adjusting the investigation probability at each step [67]. Offers a multiclass learning model that optimizes informative training compounds to support learning progress. Random Forest (RF) is used to predict quantitative compound activities. The global prediction is made by aggregating the predictions of the ensemble. Y. Brostaux [68] investigates the impact of noise in training data on the RF learning curve.

The reviewed literature focuses on improvements to known rule discovery mechanisms to transform them to become lightweight and able to be executed in a limited resource setting. In most cases, the proposed solution remains general purpose but can run with fewer required resources. Our proposal exploits the unique IoT attributes utilizing it to build a combined comprehensive framework for IoT security.

4.2 Rules generation and deployment process

The process consists of seven stages. Stage 1 composes training data from the IoT network; Stage 2 uses discovery techniques to extract essential measurements and patterns. Stage 3 consists of generating for each measure and pattern a rule. Stage 4 evaluates the effectiveness of each law against a set of training data. Stage 5 checks the generated rule set's completeness and integrity. Stage 6 simulates the same training data expecting all the designated rules to be executed. Stage 7 deploys the developed regulations set. The system is ready to accept the IoT traffic data in real-time and automatically check it against the rules set. **Figure 6** depicts The seven stages Anomaly Detection Process.

4.3 Extracting simple rules from training data

Sensor record layout includes record ID, timestamp, and values per attribute. Simple rules, such as if-then, max, min, etc., are extracted directly from the record and its associated workflows.

4.4 Compound and multi-stage rules extraction

IoT rule engines assume real-time data streaming, instant reasoning, and actuators using Machine Learning extraction of compound rules from the continuous data records. The outcome contains thresholds, measurements, and decision trees that keep expanding, consuming vast storage, memory, storage, and runtime when analyzing the decision tree for the specific rule and tracing the tree path to understand its logic. Complex Event Processing (CEP) engines support matching time-series data patterns from different sources but have downsides in IoT since the logic requires high processing power and much time. We cope with these drawbacks by reducing the number of decision trees and improving the search navigation scope to a reasonable search time. IoT attributes and functionality are used to optimize tree navigation and process sharing. We use the bootstrap aggregation technique, counting the majority vote in the

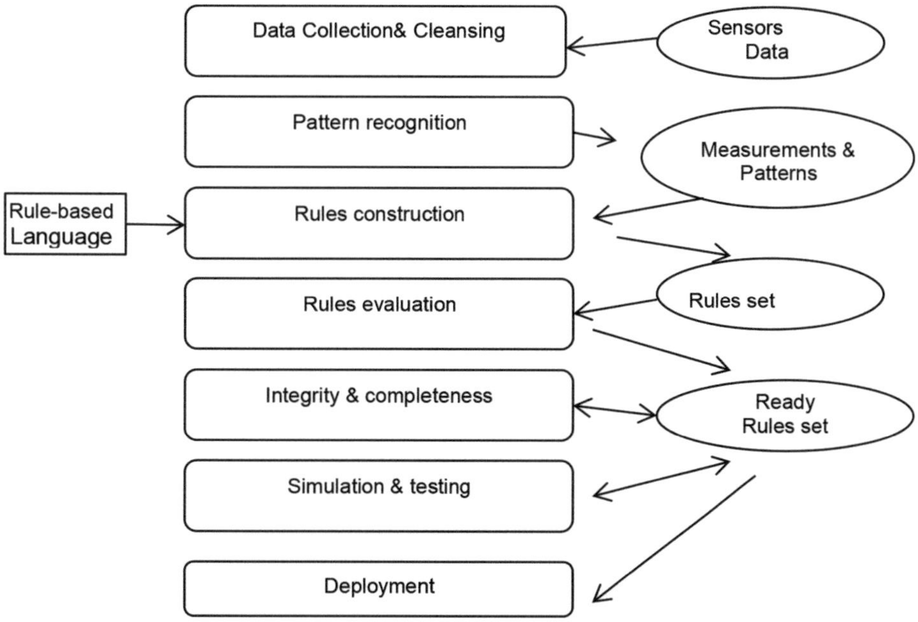

Figure 6.
The anomaly detection process.

case of decision trees. Many trees reduce the depth and width of each tree and eventually save pruning and analysis time. The algorithm accepts the number of trees, K, and the number of features, F, randomly sampled features for building a decision tree. For extensive and high-dimensional data, a large K is used. Estimating the performance of Random Forest for one core is based on the following parameters: # trees [K], # features [F], # rows [R], and maximum depth [D]. The estimated runtime formula is $K{*}F^{2}{*}R{*}2^{D}$. Hence, keeping just the most critical features, lowering the number of records, and keeping the maximum depth low will improve the overall Random Forest performance.

4.5 Experiment and summary

We use Excel functions and macros to generate compound rules such as pattern recognition for practical purposes. We also ran the Excel Machine Learning extension to create additional rules. We loaded the spreadsheet with 8 years of training data. All IoT devices are interconnected. In each device, we installed RF searching executable and deployed the generated simple rules and the RF trees in each device. We loaded the data by streaming it to the testing environment. Some generated rules do not require real-time reaction, consume processing power and memory space beyond the capacity of a typical sensor, and are executed at cloud processes. To have meaningful testing data, we intentionally added to the El-Niño file abnormal extreme values (e.g., over the maximum or lower than the minimum), wrong correlations, and classification interrupts. We loaded the data by streaming it to the testing environment. The corresponding rules and RF trees instantly detected all anomalies. We did not notice any data flowing interruptions or delays.

This section demonstrates the ability to build a lightweight, simple, and handy framework for anomaly detection, rules extraction, and rules execution given enough training data. We then described accuracy and performance improvements. Based on the accuracy and performance results, the feasibility and effectiveness of the proposed framework have been empirically proven.

5. Specific examples and case studies of successful anomaly detection

This section outlines practical and successful anomaly detection examples in various application domains. Most modern hospitals have automated laboratories, such as Chemistry, where all the blood tests are executed by dedicated machinery, which is frequently calibrated at every time interval. The calibration is done according to the manufacturer's instructions. However, some laboratory managers run ongoing anomaly detection demons to ensure real-time control. We got a request to develop an ongoing anomaly detection process that also considers actual historical testing results and incorporates an anomaly detection check that considers the history of the specific population who visited the lab in the past. We collected 3 years of lab results per machine. We ran our envelope construction process and provided a very compressed envelope considering many parameters. As a result, any machinery problem is detected in near real-time, preventing any escape of exceptional results.

Another example is detecting abnormal data streaming sequencing, timing, and frequency from a permanent external resource using a sensor for each sampled attribute. The system listens to the communication line for a while when receiving transmissions from the designated source. The method constructs a multi-dimensional envelope corresponding to each feature based on the collected features, such as timing, interval length, and frequency. The multi-dimensional envelope and a weighted compound entropy measurement provide comprehensive communications anomaly detection.

6. Limitations and practical considerations related to IoT anomaly detection mechanism

Anomaly detection systems include a preprocessing stage for defining the normal value range where any value within the specified range is designated normal. In contrast, any other value is an exception. For a time-dependent data stream, the standard value range may vary depending on the repeatable cycle, such as season or different repeatable time ranges. Therefore, the correct determination of the repeatable cycle is crucial to the accuracy of the anomaly detection process. Thus, the following vital limitations and vulnerabilities are essential to mention:

a. Identifying the repeatable cycle length is the most critical step in IoT data analysis. A wrong cycle length leads to wrong detected anomalies.

b. Detecting abnormalities at the beginning and the end of each cycle is more complex because the difference between a normal state and an abnormal state is minor; therefore, the chance of making a mistake is more significant.

c. To maintain accuracy in the standard indices, we must continuously examine the correctness of the envelope values and their adaptation to the cycle we have defined and predict natural and justified changes in the cycle and its corresponding values used to check the anomalies over time.

7. Conclusion

This chapter deals with current and future trends in Anomaly detection concepts and technologies for the IoT context. We started with an overview of various IoT applications spread over most functional domains, such as Industry machinery, Health, Smart home, and smart city. Most of the new developments in IoT focus on solutions to the severe security breach caused by interconnecting numerous IoT devices to the Internet. These solutions provide tools for detecting/identifying operations anomalies. Therefore, we allocated Section 2 to cover IoT operation and communications security aspects. Then we elaborated on generating an envelope for anomaly detection for temporal transactions, which are the nature of IoT activity and networks. We finally elaborate on advanced technology for anomaly detection using Random Forest distributed over a network of IoT devices.

IoT keeps evolving and spreading fast everywhere in all functional domains in the modern world. Thus, new developments and recent trends will continue growing, so new chapters will follow.

Author details

Menachem Domb[1*], Sujata Joshi[2] and Arulmozhi Khn[2]

1 Ashkelon Academy College [AAC], Ashkelon, Israel

2 Symbiosis Institute of Digital and Telecom Management, Symbiosis International (Deemed University), Pune, India

*Address all correspondence to: dombmnc@edu.aac.ac.il

References

[1] Chatterjee A, Ahmed BS. IoT anomaly detection methods and applications (survey). Internet of Things. 2022;**19**:100568. DOI: 10.1016/j.iot.2022.100568

[2] Chalapathy R, Chawla S. Deep learning for anomaly detection: A survey. 2019. arXiv:1901.03407 Google Scholar

[3] Himeur Y, Ghanem K, Alsalemi A, Bensaali F, Amira A. Artificial intelligence based anomaly detection of energy consumption in buildings: A review, current trends, and new perspectives. Applied Energy. Elsevier; 2021;**287**:1-26. Article 116601. DOI: 10.1016/j.apenergy.2021.116601. Available from: https://www.sciencedirect.com/science/article/pii/S0306261921001409

[4] Talagala PD, Hyndman RJ, Smith-Miles K. Anomaly detection in high-dimensional data. Journal of Computational and Graphical Statistics. 2021;**30**(2):360-374. DOI: 10.1080/10618600.2020.1807997

[5] Yin C, Zhang S, Wang J, Xiong NN. Anomaly detection based on convolutional recurrent auto-encoder for IoT time series. IEEE Transactions on Systems, Man, and Cybernetics: Systems. 2022;**52**(1):112-122. DOI: 10.1109/TSMC.2020.2968516

[6] Sobhani M, Hong T, Martin C. Temperature anomaly detection for electric load forecasting. International Journal of Forecasting. 2020; 36 (2): 324-333. DOI: 10.1016/j.ijforecast.2019.04.022. Available from: https://www.sciencedirect.com/science/article/pii/S0169207019301633

[7] Asakura T, Yashima W, Suzuki K, Shimotou M. Anomaly detection in a logistic operating system using the Mahalanobis–Taguchi method. Applied Sciences. Basel, Switzerland: MDPI; 2020;**10**(12):1-25. DOI: 10.3390/app10124376. Available from: https://www.mdpi.com/2076-3417/10/12/4376

[8] Huang S, Guo Y, Yang N, Zha S, Liu D, Fang W. A weighted fuzzy C-means clustering method with density peak for anomaly detection in IoT-enabled manufacturing process. Journal of Intelligent Manufacturing. Germany: Springer; 2021;**32**:1845-1861. DOI: 10.1007/s10845-020-01690-y

[9] Yasaei R, Hernandez F, Al Faruque MA. IoT-CAD: Context-aware adaptive anomaly detection in IoT systems through sensor association. In: 2020 IEEE/ACM International Conference on Computer-Aided Design, ICCAD. NY, USA: ACM; 2020. pp. 1-9

[10] Zekry A, Sayed A, Moussa M, Elhabiby M. Anomaly detection using IoT sensor-assisted ConvLSTM models for connected vehicles. In: 2021 IEEE 93rd Vehicular Technology Conference, VTC2021-Spring. New York, USA: IEEE; 2021. pp. 1-6. DOI: 10.1109/VTC2021-Spring51267.2021.9449086

[11] Wang J, Tang Y, He S, Zhao C, Sharma PK, Alfarraj O, et al. LogEvent2vec: LogEvent-to-vector based anomaly detection for large-scale logs in the Internet of Things. Sensors. Basel, Switzerland: MDPI; 2020;**20**(9):1-27. DOI: 10.3390/s20092451. Available from: https://www.mdpi.com/1424-8220/20/9/2451

[12] Xu R, Cheng Y, Liu Z, Xie Y, Yang Y. Improved long short-term memory (LSTM) based anomaly detection with concept drift adaptive method for supporting IoT services. Future

Generation Computer Systems. 2020; 112: 228-242. DOI: 10.1016/j.future.2020.05.035. Available from: https://www.sciencedirect.com/science/article/pii/S0167739X20302235

[13] Tripathi G, Abdul Ahad M, Paiva S. SMS: A secure healthcare model for smart cities. Electronics. Basel, Switzerland: MDPI; 2020;**9**(7):1-18. DOI: 10.3390/ electronics9071135. Available from: https://www.mdpi.com/2079-9292/9/7/1135

[14] Ullah W, Ullah A, Haq IU, Muhammad K, Sajjad M, Baik SW. CNN features with bi-directional LSTM for real-time anomaly detection in surveillance networks. Multimedia Tools and Applications. 2021;**80**(11):16979-16995

[15] Tsogbaatar E, Bhuyan MH, Tanaka Y, Fall D, Gonchigsumlaa K, Elmroth E, et al. Del-IoT: A deep ensemble learning approach to uncover anomalies in IoT, Internet of Things. 2021;**14**:100391. DOI: 10.1016/j.iot.2021.100391. Available from: https://www.sciencedirect.com/science/article/pii/S2542660521000354

[16] Mirsky Y, Golomb T, Elovici Y. Lightweight collaborative anomaly detection for the IoT using blockchain. Journal of Parallel and Distributed Computing. 2020;**145**:75-97. DOI: 10.1016/j.jpdc.2020.06.008. Available from: https://www.sciencedirect.com/science/article/pii/S0743731520303154

[17] An Y, Yu FR, Li J, Chen J, Leung VCM. Edge intelligence (EI)-enabled HTTP anomaly detection framework for the Internet of things (IoT). IEEE Internet of Things Journal. 2021;**8**(5):3554-3566. DOI: 10.1109/JIOT.2020.3024645

[18] Shen M, Tang X, Zhu L, Du X, Guizani M. Privacy-preserving support vector machine training over blockchain-based encrypted IoT data in smart cities. IEEE Internet of Things Journal. 2019;**6**(5):7702-7712. DOI: 10.1109/JIOT.2019.2901840

[19] Wan J et al. Software defined industrial IoT in the context of industry 4.0. IEEE Sensors Journal. 2016;**16**(20):7373-7380. DOI: 10.1109/JSEN.2016.2565621

[20] Lemayian JP, Al-Turjman F. Intelligent IoT communication in smart environments: An overview. In: Artificial Intelligence in IoT. Transactions on Computational Science and Computational Intelligence. Singapore: Springer; 2019. DOI: 10.1007/978-3-030-04110-6_10

[21] Wang KH, Chen CM, Fang W, Wu TY. A new ultra-lightweight authentication protocol in IoT environment for RFID tags. The Journal of Supercomputing. 2018;**74**(1):65-70. DOI: 10.1007/s11227-017-2105-8

[22] Singh S, Sharma PK, Moon SY, Park JH. Advanced lightweight encryption algorithms for IoT devices: Survey, challenges and solutions. Journal of Ambient Intelligence and Humanized Computing. Germany: Springer; 2017;**18**:1. DOI: 10.1007/s12652-017-0494-4

[23] Rachit SB, Ragiri PR. Security trends in Internet of Things: A survey. SN Applied Sciences. 2021;**3**(1):1-14. DOI: 10.1007/s42452-021-04156-9

[24] Bembe M, Abu-Mahfouz A, Masonta M, Ngqondi T. A survey on low-power wide area networks for IoT applications. Telecommunication Systems. 2019;**71**(2):249-274. DOI: 10.1007/s11235-019-00557-9

[25] Zhang X, Wen F. A novel anonymous user WSN authentication for Internet of Things. Soft Computing.

2019;**23**(14):5683-5691. DOI: 10.1007/s00500-018-3226-6

[26] Alshehri MD, Hussain FK. A fuzzy security protocol for trust management in the Internet of things (Fuzzy-IoT). Computing. 2019;**101**(7):791-818. DOI: 10.1007/s00607-018-0685-7

[27] Chen H, Meng C, Shan Z, Fu Z, Bhargava BK. A novel low-rate denial of service attack detection approach in Zigbee wireless sensor network by combining Hilbert-Huang transformation and trust evaluation. IEEE Access. 2019;**7**:32853-32866. DOI: 10.1109/ACCESS.2019.2903816

[28] Gubbi J, Palaniswami M, Buyya R, Marusic S. Internet of Things: A vision, architectural elements, and future directions. Future Generation Computer Systems. 2013;**29**(7):1645-1660. DOI: 10.1016/j.future.2013.01.010

[29] Li S, Da Xu L, Zhao S. 5G Internet of Things: A survey. Journal of Industrial Information Integration. 2018;**10**:1-9. DOI: 10.1016/j.jii.2018.01.005

[30] Arfaoui G et al. A security architecture for 5G networks. IEEE Access. 2018;**6**:22466-22479. DOI: 10.1109/ACCESS.2018.2827419

[31] Mohanty SN et al. An efficient lightweight integrated blockchain (ELIB) model for IoT security and privacy. Future Generation Computer Systems. 2020;**102**:1027-1037. DOI: 10.1016/j.future.2019.09.050

[32] Chatterjee S, Mukherjee R, Ghosh S, Ghosh D, Ghosh S, Mukherjee A. Internet of Things and cognitive radio - Issues and challenges. In: 2017 4th International Conference on Opto-Electronics and Applied Optics (Optronix) 2017. NY, USA: IEEE; 2018. pp. 1-4. DOI: 10.1109/OPTRONIX.2017.8349993

[33] Fortino G, Russo W, Savaglio C. Simulation of agent-oriented Internet of things systems. In: CEUR Workshop Proc. Vol. 1664. 2016. pp. 8-13

[34] Leloglu E. A review of security concerns in the Internet of Things. Journal of Communications and Computers. 2017;**5**(01):121-136. DOI: 10.4236/jcc.2017.51010

[35] Goyal P, Sahoo AK, Sharma TK. Internet of things: Architecture and enabling technologies. Materials Today: Proceedings. 2019;**34**(January):719-735. DOI: 10.1016/j.matpr.2020.04.678

[36] Soni A, Upadhyay R, Jain A. Internet of Things and Wireless Physical Layer Security: A Survey. In: Computer Communication, Networking and Internet Security: Proceedings of IC3T. Singapore: Springer; 2017. pp. 115-123. DOI: 10.1007/978-981-10-3226-4_11

[37] Xu H, Sgandurra D, Mayes K, Li P, Wang R. Analyzing the resilience of the Internet of things against physical and proximity attacks. Security, Privacy, and Anonymity in Computation, Communication, and Storage: SpaCCS 2017 International Workshops, Guangzhou, China; Switzerland. In: Proceedings 10. In: Lect. Notes Computer Science. (including Subser. Lect. Notes Bioinformatics), 12-15 December 2017. Switzerland: Springer International Publishing; Vol. 10658 LNCS. 2017. pp. 291-301. DOI: 10.1007/978-3-319-72395-2_27

[38] Salim MM, Rathore S, Park JH. Distributed denial of service attacks and its defenses in IoT: A survey. Vol. 76(7). US: Springer; 2020. DOI: 10.1007/s11227-019-02945-z

[39] Stiawan D, Idris MY, Malik RF, Nurmaini S, Alsharif N, Budiarto R. Investigating Brute force attack patterns

in IoT network. Journal of Electrical and Computer Engineering. Hindawi; 2019;**2019**:1-14. DOI: 10.1155/2019/4568368

[40] Shen H, Shen J, Khan MK, Lee JH. Efficient RFID authentication using elliptic curve cryptography for the Internet of Things. Wireless Personal Communications. 2017;**96**(4):5253-5266. DOI: 10.1007/s11277-016-3739-1

[41] Om Kumar CU, Sathia Bhama PRK. Detecting and confronting flash attacks from IoT botnets. The Journal of Supercomputing. 2019;**75**(12):8312-8338. DOI: 10.1007/s11227-019-03005-2

[42] Mairal J, Ponce J, Bach F, Sapiro G. Online dictionary learning for sparse coding. In: 26th Annual International Conference on Machine Learning. NY, USA: ACM; 2009. pp. 689-696

[43] Dietterich TG. Machine Learning for Sequential Data, Joint IAPR and Structural and Syntactic Pattern Recognition (SSPR). Germany: Springer; 2002. pp. 15-30

[44] Vlachos M, Freris NM, Kyrillidis A. Compressive mining: Fast and optimal data mining in the compressed domain. The VLDB Journal. 2015;**24**(1):1-24

[45] Sakurada M, Yairi T. Anomaly detection using autoencoders nonlinear dimensional reduction, MLSDA 2014. In: Machine Learning for Sensory Data Analysis. NY, USA: ACM; 2014. pp. 4-11. DOI: 10.1145/2689746.2689747

[46] Reeves G, Liu J, Nath S, Zhao F. Managing massive time series streams with multi-scale compressed trickles. Proceedings of the VLDB Endowment. 2009;**2**(1):97-108

[47] Chilimbi TM, Hirzel M. Dynamic hot data stream prefetching for general purpose programs. In: ACM SIGPLAN Notices. Vol. 37(5). NY USA: ACM; 2002. pp. 199-209

[48] Lane T, Brodley CE. Temporal sequence learning and data reduction for anomaly detection. ACM TISSEC. 1999;**2**(3):295-331

[49] Kasiviswanathan SP, Melville P, Banerjee A, Sindhwani V. Emerging topic detection using dictionary learning. In: Proceedings of the 20th ACM international conference on Information and knowledge management. NY, USA: ACM; 2011. pp. 745-754

[50] Aldroubi A, Cabrelli C, Molter U. Optimal nonlinear models for sparsity and sampling. Journal of Fourier Analysis and Applications. 2008;**14**(5-6):793-812

[51] Rubinstein R, Bruckstein AM, Elad M. Dictionaries for sparse representation modeling. Proceedings of the IEEE. 2010;**98**(6):1045-1057

[52] Cherian A, Sra S, Papanikolopoulos N. Denoising sparse noise via online dictionary learning. In: IEEE International Conference on Acoustics, Speech and Signal Processing (ICASSP). NY, USA: IEEE; 2011. pp. 2060-2063

[53] Duarte-Carvajalino JM, Sapiro G. Learning to sense sparse signals: Simultaneous sensing matrix and sparsifying dictionary optimization, DTIC Document, Tech. Rep. 2008

[54] Mansoori EG, Zolghadri MJ, Katebi SD. SGERD: A steady-state genetic algorithm for extracting fuzzy classification rules from data. IEEE Transactions of Fuzzy Systems. 2008;**16**(4):1061-1071 ISSN: 1063-6706

[55] Extracting decision rules from qualitative data using Sugeno integral.

In: Proceedings of the 13th European Conference, ECSQARU 2015, Compiègne, France. July 2015; Vol. 9161. pp. 14-24. ISBN 978-3-319-20806-0. ISSN 0302-9743

[56] Daniel B, Gulwani S, Hart T, Zorn B. FlashRelate: extracting relational data from semi-structured spreadsheets using examples, Proceedings of the 36th ACM SIGPLAN Conference on Programming Language Design and Implementation. New York: ACM. June 2015; Vol. 50(6). pp. 218-228

[57] Very fast decision rules for classification in data streams, data mining and knowledge discovery. 2015;**29**(1):168-202 ISSN1384-5810

[58] Jafarzadeh H, Torkashvand RR, Asgari C, Amiry A. Provide a new approach for mining fuzzy association rules using apriori algorithm. Indian Journal of Science and Technology. 2015;**8**(S7):127-134 ISSN: 0974-6846

[59] Pourpanaha F, Limb CP, Saleh JM. A hybrid model of fuzzy ARTMAP and genetic algorithm for data classification and rule extraction. Expert Systems with Applications. 2016;**49**(7):4-85

[60] Mashinchi R, Selamat A, Ibrahim S, Krejcar O. Granular-Rule Extraction to Simplify Data. In: Nguyen N, Trawiński B, Kosala R, editors. Intelligent Information and Database Systems. ACIIDS 2015. Lecture Notes in Computer Science. vol. 9012. Germany, Cham: Springer; 2015. pp. 421-429. DOI: 10.1007/978-3-319-15705-4_41

[61] Yang H, Xiao C, Qiao Y. Study on anomaly detection algorithm of QAR data based on attribute support of rough set. International Journal of Hybrid Information Technology. 2015;**8**(1):371-382 ISSN: 1738-9968

[62] Tang H. A simple approach of data mining in excel. In: 2008 4th International Conference on Wireless Communications, Networking and Mobile Computing, Dalian, China. Piscataway, NJ, USA: IEEExplore; 2008. pp. 1-4. DOI: 10.1109/WiCom.2008.2679

[63] Tong S, Koller D. Support Vector Machine Active Learning with Applications to Text Classification. Journal of Machine Learning Research. NY, USA: Microtom Publishing; 2001;**2**(1):45-66. DOI: 10.1162/153244302760185243

[64] Rasmussen CE. Support Vector Machine Active Learning with Applications to Text Classification. CiteSeerX; 2006

[65] Osugi T, Kim D, Scott S. Balancing Exploration and Exploitation: A New Algorithm for Active Machine Learning. In: 5th IEEE International Conference on Data Mining. NY, USA: IEEE; 2005. pp. 8. DOI: 10.1109/ICDM.2005.33

[66] Lang T, Flachsenberg F, von Luxburg U, Rarey M. Feasibility of active machine learning for multiclass compound classification. 2016. DOI: 10.1021/acs.jcim.5b00332

[67] Trees SB, Jothi Venkataeswaran C. Improving classification accuracy based on random forest model with uncorrelated high performing. International Journal of Computer Applications. 2014;**101**:(13)

[68] Brostaux Y. Random forests and decision trees clasifiers effects of data quality on the learning curve, ibs2006_poster

Chapter 4

Anomaly Detection in Time Series: Current Focus and Future Challenges

Farrukh Arslan, Aqib Javaid, Muhammad Danish Zaheer Awan and Ebad-ur-Rehman

Abstract

Anomaly detection in time series has become an increasingly vital task, with applications such as fraud detection and intrusion monitoring. Tackling this problem requires an array of approaches, including statistical analysis, machine learning, and deep learning. Various techniques have been proposed to cater to the complexity of this problem. However, there are still numerous challenges in the field concerning how best to process high-dimensional and complex data streams in real time. This chapter offers insight into the cutting-edge models for anomaly detection in time series. Several of the models are discussed and their advantages and disadvantages are explored. We also look at new areas of research that are being explored by researchers today as their current focuses and how those new models or techniques are being implemented in them as they try to solve unique problems posed by complex data, high-volume data streams, and a need for real-time processing. These research areas will provide concrete examples of the applications of discussed models. Lastly, we identify some of the current issues and suggest future directions for research concerning anomaly detection systems. We aim to provide readers with a comprehensive picture of what is already out there so they can better understand the space – preparing them for further development within this growing field.

Keywords: anomaly detection, anomaly detection in time series, high dimensional data, big data, current focus and future challenges, machine learning, deep learning, forecasting, real time

1. Introduction

Time series data mining is becoming increasingly important due to the advances in technology which have allowed us to collect and store large amounts of structured temporal data. A wide range of tasks can be performed with this time series data such as classification [1], clustering [2], forecasting [3] and outlier detection [4–6]. Extracting meaningful insights from this data opens up new opportunities for research across diverse areas.

Anomaly detection in time series is a critical task with significant implications in numerous fields, including finance [7], healthcare [8], and security [9]. Identifying and analyzing outliers in time-series data is a critically important operation for obtaining meaningful insights [6]. As described in a seminal paper, there are two types of univariate time-series outliers - type I and type II [10, 11]. Whilst type I outlier events occur individually, type II can be linked to subsequent power shifts. It's essential to have an in-depth understanding of both these outlier types if a meaningful analysis of the data is to be undertaken.

The detection of unusual patterns or events within data streams can provide valuable insights and help identify potentially harmful or fraudulent behavior. However, processing high-dimensional [12, 13] and complex data streams [14] in real-time [15] remains a challenging task. In recent years, many statistical [16], machine learning [17], and deep learning [18] techniques have been proposed to tackle these challenges.

This chapter provides an overview of the current state-of-the-art models for anomaly detection in time series, including their strengths and limitations. We also explore new areas of research as current focuses of researchers in this field, that are being explored by implementing recent techniques, models of machine and deep learning to solve unique challenges posed by high-dimensional and complex data, high-volume data streams, and the need for real-time processing.

One of the primary challenges of anomaly detection in time series is dealing with high-dimensional and complex data. Traditional statistical methods such as ARIMA [19] and exponential smoothing [20] have been used in the past but lack the flexibility to handle complex data with multiple attributes. Newer techniques, including machine learning-based approaches such as isolation forests [21], autoencoder-based methods [22], and deep learning techniques such as LSTM and CNNs [18], have shown promising results in detecting anomalies in high-dimensional and complex data. However, these approaches also have their limitations, including high computational costs and the need for extensive training data [23].

Another challenge in anomaly detection in time series is dealing with high-volume data streams in real-time [24]. Traditional batch processing methods are not suitable for real-time applications where timely detection is critical. As a result, new methods such as online anomaly detection algorithms [25] and sliding window-based [26] approaches have been proposed, which process data in a continuous and efficient manner.

In addition to the current state-of-the-art models, techniques, and areas of research, this chapter also identifies some of the current issues and suggests future directions for research concerning anomaly detection systems. These include the need for more interpretable models, the development of novel unsupervised methods, the integration of domain knowledge, and the need to address the issue of data imbalance.

Sections of this chapter discuss recent algorithms, models separately as well as within current wide areas of research i.e., Forecasting, Real time anomaly detection, Dealing with Big and high dimensional data processing problem, Anomaly detection using Artificial Intelligence (AI), Industrial Control Systems. There will be a literature review of many models and strategies implemented in these areas. All these areas as well as models will be solving challenges as discussed before. Overall, the purpose is to provide readers a literature review about current focuses of researchers and to facilitate further development and research in time series data anomaly detection.

2. Big data and high dimensionality

Modern data sets are increasingly high-volume, high-velocity, and high-variety, making it difficult to identify anomalies with accuracy. Researchers and developers have started to investigate new approaches for coping with this complexity. When the number of features grows exponentially, more data is needed to create accurate models - leading to sparse and isolated data points. Gartner [13] defines big data as a collection of attributes that require cost-effective analytics to generate insight. The key challenges associated with big data are described in the 5 Vs: value, veracity, variety, velocity and volume [27].

The next stated paragraphs will give an overview of how real-time big data processing can be used to detect anomalous events through machine learning algorithms, as well as its current limitations and challenges.

McNeil et al. [28] examined existing tools used to detect malware on mobile devices. It was noted that these methods lacked the capability to incorporate group user profiling, which is necessary to automate behavior-based dynamic analysis for focused malware identification. To overcome this, they demonstrated a scalable architecture called SCREDENT, which allows users to classify, identify, and forecast possible target malwares in real time. While initial evaluation indicated that the approach had promise, further testing failed to demonstrate desired results.

In reference [29], a new architecture was proposed to detect threats in real-time using stream processing and machine learning. This architecture promotes an environment with minimal human oversight, allowing for improved detection of both known and previously unseen cyberattacks in order to hone attack classification and anomaly detection capabilities. However, their results did not benefit from the open KDD dataset as much as expected.

In reference [30], complex network infrastructure with vast logs files became the focus of another approach which looks to assess security logs that contain various device information through data mining and machine learning techniques. This proposed approach is split into two phases: defining/configuring the detection mechanism and then executing it at runtime. Nevertheless, the practical implementation turned out to need more automation due to its high human intervention levels; similarly, its output accuracy was not precise enough.

Recent research has highlighted the use of machine learning models for anomaly detection. But, the inability of inherent system performance to keep up with increasing network traffic is an issue that needs to be addressed. To do so, a novel model utilizing Hadoop, HDFS, MapReduce, cloud and multiple machine learning algorithms were developed. Weka interface was used to assess accuracy and efficiency through naive bayes, decision tree and support vector methods. However, the implementation of cloud infrastructure and real-time data streaming has not been sufficiently discussed as yet in this project [31].

Research paper [32] focuses on anomaly detection in streaming data, and provides a new approach to evaluate online anomaly detection with entropy and Pearson correlation. Big data streaming components like Kafka queues and Spark Streaming are used as a means of ensuring scalability and generality, although some processes which were potentially complicated by long batch processing periods or data limitations were not resolved.

Researchers have proposed a method for anomaly detection in smart grids that uses real-time, minimal energy consumption [33]. The proposed in-memory distributed framework, comprised of Spark Streaming and Lambda System, is viable for scalable live streaming. However, it did take longer to train the model and there were scheduling issues with real-time tasks.

In reference [34], another framework was presented - one which involves sensor data preprocessing: anomaly detection using principal statistical analysis and Bayesian networks; as well as sensor data redundancy elimination using static or dynamic Bayesian networks (SBNs/DBNs). Included were two algorithms: static sensor data redundancy detection algorithm (SSDRDA) and real-time sensor data redundancy detection algorithm (RSDRDA), both serving to reduce redundant data in either static datasets or real-time scenarios respectively.

Anomaly detection in time series requires modifications in the above framework of approaches for efficiency. Anomaly detection in real-time big data analytics is a promising area of study, particularly when machine learning techniques are incorporated. Advancements in this field are likely to yield high accuracy and efficiency. Thus, the potential benefits of such research cannot be underestimated.

Following points will lay out the foremost research challenges in this field, in order to promote progress.

Redundancy: Managing real-time big data from diverse sources is challenging. Current technologies like Hadoop and spark fail to address redundancy, data quality and reliability, cost [35], and storage schema [36]. A new framework is needed to tackle these complexities.

Computational cost: Anomaly detection requires multiple techniques, increasing computation cost. Large datasets and high dimensionality cause algorithmic instability and computational expense [23]. Big data and cloud technology enable parallel and distributed processing and reduce computing costs. Cheaper processors and high chips improve system power and data processing in real-time, minimizing computational expenses.

Nature of Input data: Input data has instances with binary/categorical or continuous attributes, and can be univariate or multivariate. Anomaly detection algorithm selection depends on data diversity and attribute type [37]. A hybrid framework using unsupervised machine learning algorithms can detect anomalies in different datasets.

Noise and missing value: Network sensor streaming data has various types and can produce false alarms due to noise and missing values from high speed [37]. Noise can hide true anomalies [38]. An auto noise cleansing module in the detection framework can remove unnecessary features and handle missing values.

Parameters Selection: Optimal parameters for machine learning algorithms are hard to select [39]. Real-time anomaly detector needs to consider multiple and single hyperparameters, which may change over time [40]. Parameter choice affects algorithm performance; eccentricity techniques can reduce selection processes [41].

Inadequate Architecture: Organizations need big data architecture for large real-time data. Existing architectures are insufficient. Real-time analytics and application components can create efficient environment [42]. Big data technologies and hybrid machine learning algorithms can solve architectural problems. Scalability for data in motion and at rest is achieved.

Data visualizations: Data or reports need effective, visual insights. Anomalies from connected devices can use heat maps, scatter plots, parallel coordinates and node-link graphs for 2D/3D views. 3D interaction needs data understanding and user rotation and zoom [43]. Opensource visualization techniques in frameworks can automatically select techniques for better user experience.

Heterogeneity of data: Unstructured data is varied and large, such as emails, faxes, form documents, social media posts, etc. Transcription is expensive. Hybrid

Machine Learning algorithms can identify data types quickly and accurately. Complex machine learning models can recognize heterogeneous information sources from unstructured text.

Accuracy: Anomaly detection with existing technologies is inaccurate. A hybrid machine learning algorithm can analyze large data from modern applications with low memory and power. Our team combines real time big data technologies with this algorithm for efficient and accurate results.

3. Transformer

The Transformer architecture, introduced by Vaswani et al. [44], has been widely used in various natural language processing tasks, such as machine translation and text classification. However, its effectiveness in anomaly detection is limited due to the lack of a specific mechanism to capture anomalous patterns. To address the shortcoming of Transformer architecture, researchers have created a variety of modifications to attempt to improve its performance, including the incorporation of an Anomaly-Attention mechanism. One such modification is the Anomaly Transformer, as illustrated in reference [45], which utilizes this mechanism to improve the detection of anomalous patterns in data. Thus, the Anomaly Transformer architecture represents an important advancement in the application of Transformers for anomaly detection.

Recent studies have focused on leveraging Transformer based architectures to benefit the time series anomaly detection task. These approaches are capable of modeling temporal dependencies, enabling better anomaly detection quality [45]. For instance, TranAD [46], MT-RVAE [47] and TransAnomaly [48] all fuse Transformers with VAEs i.e., neural generative models [49] or GANs [50], demonstrating improved performance in anomaly detection. These models have been explained further below.

TranAD: TranAD [46] is a robust adversarial training procedure designed to address small deviations in anomalies that the typical Transformer-based network may overlook. This GAN-style approach consists of two transformer encoders and two decoders in order to maintain stability. The results of an ablation study illustrate the performance of this architecture, with F1 scores dropping by nearly 11% when the Transformer-based encoder-decoder was replaced. This clearly demonstrates the efficacy of using Transformers for time series anomaly detection. It's valuable for modern industrial systems where instant detection of anomalies is compulsory.

MT-RVAE: There are two different approaches that combine VAE and Transformer to create novel models for time-series analysis. MT-RVAE, proposed by Wang et al. [47], uses a multiscale Transformer model to extract information from sequences of varying scales like complex satellite systems with several subsystems. Each subsystem's temporal features must be analyzed in correlation [47]. This addresses the limitations traditional Transformers had in being able to accurately analyze sequential data as these models were limited to local information extraction only. TransAnomaly, proposed by Zhang et al. [48], combines VAE with transformer for increased parallelization purposes. The combination of these two techniques is predicted to reduce training costs up to 80%.

GTA: GTA [51] leverages Transformers and graph-based learning to accurately detect anomalies in multivariate time series data, even when there are few dimensions or limited close relationships among sequences. This method features a multi-branch attention mechanism composed of global-learned attention, regular multi-head attention, and neighborhood convolution for increased accuracy, as well as a graph convolution structure for modeling influence propagation processes. Thus, GTA seeks to provide an improved approach for analyzing and detecting anomalies in multivariate time series data than previous methods. It's valuable for internet connected sensory devices like smart power grids, water distribution networks as they remain under attack of cyber-attacks [51].

AnomalyTrans: AnomalyTrans [45] is a novel approach in distinguishing anomalies. Drawing inspiration from TranAD, AnomalyTrans makes it more difficult for anomalies to create strong connections with the entire time series, though retaining connectivity between adjacent time points. The model leverages Transformer and Gaussian prior-association to reach this objective. Through utilizing a minimax strategy to optimize the anomaly model, AnomalyTrans enforces restrictions on prior- and series-associations that result in a greater divergence between them.

D3TN: Disentangled Dynamic Deviation Transformer Network (D3TN) [52] is highly effective system for multivariate time series anomaly detection. It considers both short-term and long-term temporal dependencies as well as complex inter-sensor dependencies. To better model static topology, a new disentangled multi-scale aggregation scheme for graph convolutional neurons for fixed inter-sensor relationships was introduced. A self-attention mechanism was also employed to capture dynamic directed interaction in various subspaces that vary with time and unexpected events. Moreover, parallel processing of the time series helps simulate complex temporal correlations that span multiple time periods.

DATN: The Decompositional Auto-Transformer Network (DATN) [53] is a unique anomaly detection method for time series. This novel approach breaks complex time series into seasonal and trend components, before then renovating them with deep models. Additionally, the design integrates an auto-transfomer block to detect important representations and dependencies based on seasonality and trends in the series. Furthermore, rather than using a traditional complex transformer decoder, we substitute it with a more efficient linear decoder.

Transformers have been applied in many real word scenarios for anomaly detection in time series like: SMD is a 5-week-long dataset acquired from one of the leading Internet companies with 38 characteristics. Pooled Server Metrics (PSM) was procured internally from multiple server nodes at eBay and consists of 26 variables. Besides these, both Mars Science Laboratory rover (MSL) and Soil Moisture Active Passive (SMAP) satellite datasets from NASA have been compiled as well, containing 55 and 25 features respectively with regard to the anomaly data derived from the Incident Surprise Anomaly (ISA) reports for spacecraft monitoring systems. Last but not least, the Secure Water Treatment (SWaT) dataset contains 51 indicators derived from continuous operation on a critical infrastructure system [45, 53].

The possible future challenges are indicated below in paragraphs.

Inductive Biases for Time Series Transformers: Transformers are powerful, general networks for modeling long-range dependencies. But they require quite a bit of data to train effectively and avoid falling prey to data overfitting. Time series data often follows seasonal or periodic patterns, as well as other trends, which suggests that incorporating this information into Transformers has potential to lead improvement in performance. For instance, recent studies have demonstrated the effectiveness of frequency processing [54] and capturing series periodicity [55]. Additionally, both explicitly allowing cross-channel dependency [56] and preventing it via channel-independent attention module [57] have yielded better models for certain tasks. The challenge then lies in finding a balance between designing inductive bias to suppress noise while amplifying signal — a task whose solution is yet to come but promises exciting possibilities ahead.

Transformers and GNN for Time Series: As datasets with multi-dimensional and spatial–temporal characteristics become more widespread, it's essential to have tools which can effectively capture the complexities that these data represent. Graph Neural Networks (GNNs) is one method of modeling dependencies and relations with each other between dimensions. Recent studies have shown that combining GNNs with Transformers/Attentions leads to impressive performance improvement in areas such as traffic forecasting [58, 59] and multi-modal forecasting [60, 61], knowledge of latent causality and the underlying clarity of spatial–temporal performance can be increased with a greater comprehension. It is an important development that could result in more effective use of Transformer-GNN hybrid models for spatial–temporal modeling in time series going forward.

Pre-trained Transformers: As the advances of large-scale Transformers using pre-training have yielded observable improvements across a wide range of natural language processing tasks [62, 63] and CV [64], research conducted on their efficacy for time series applications has been limited. Works existing to this day primarily focus on classification activities [65, 66]. In order to develop effective pre-trained Transformer models that are equipped to address a range of use cases within time series analysis, further examination will be required in the future.

Architecture Level Variants: Considering the success of Transformer variants in NLP and CV, it may be beneficial to transfer this concept over to time series data and tasks. We can look into more architecture-level designs for Transformers which may optimize performance on time series specific models. Examples of these variants include lightweight [67, 68], cross-block connectivity [69], adaptive computation time [70, 71], and recurrence [72]. These architecture-level designs provide us with a whole new range of opportunities for improvement.

Transformers with Neural Architecture Search: Tuning Transformer hyper-parameters such as embedding dimension, heads, and layers can have a significant impact on performance. Thankfully, Neural Architecture Search (NAS) provides a means to automatically find architectures that optimize performance. Recently, NAS technologies in NLP and CV have been applied to transformers [73, 74]. For machine data which may be high dimensional yet long in length, this technique is especially important for designing memory- and computationally-efficient transformers. We anticipate further progress in this area as the industry gears up for more efficient time series Transformers.

4. Non-pattern anomaly detection

Non-Pattern Anomaly Detection is an underdiagnosed but powerful method of identifying anomalies in time series. Existing techniques use initial profiling to determine which behavior should be tagged as "normal" or "abnormal," but this definition fails to capture the nuanced changes between situations in different conditions. Researchers recognized the importance of such a technique and emphasized its potential for detecting abnormalities even in the absence of statistical methods that often play a dominant role in machine learning processes. Team of researchers aimed to compare current machine learning algorithms relating to NP-AD approaches and assessed how various datasets demonstrated their capacity for anomalies on diverse situations [75].

5. Hybrid models

Multivariate time-series anomaly detection is a complex challenge due to the imbalance of anomalous data and its underlying intricacies. Combining different methods for detecting anomalies in time series has been well explored, resulting in improved accuracy. Notably, hybrid models combining statistical and deep learning approaches have been found to provide greater precision when determining uncertainty and quantifying forecasts associated with these models.

One example hybrid approach is called HAD-MDGAT – it's based on a GAT (graph attention network) combined with multi-channel temporal stacked Denoising Autoencoder (MDA), designed to learn temporal and spatial correlations among observations. Ablation study results show that MDA enhances anomaly detection accuracy dramatically; this model with an MDA layer scored 10.86% higher than one without the extra layer [76].

A research paper published in reference [77] outlined a novel Long Short-Term Memory (LSTM) network-based method for accurately forecasting multivariate time series data. In addition, the study featured an LSTM Autoencoder network-based approach coupled with a one-class Support Vector Machine (SVM) algorithm, which was employed for anomaly detection. Their findings demonstrated that the LSTM Autoencoder based method outperforms the previously proposed LSTM based method. Moreover, their proposed forecast approach surpassed several other methods by NASA. The LSTM based methodology is well suited to forecasting while the combination of an LSTM Autoencoder with the OCSVM is suitable for detecting anomalies [77].

MES-LSTM is a combination of a multivariate forecasting model and Long Short-Term Memory, a form of Recurrent Neural Network (RNN). Accurate attribution is an important part of any system as it reinforces confidence in the mechanics and makes sure learning processes are not based on spurious effects. While MES-LSTM does a great job of anomaly detection, overall performance could still benefit from improvement [78].

A hybrid deep-learning model that integrates long short-term memory (LSTM) and autoencoder (AE) networks was proposed for anomaly detection tasks in Indoor Air Quality (IAQ) time series data. The LSTM cells are stacked together to learn the long-term dependencies in time-series data, while the AE helps identify an optimal threshold based on reconstruction loss rates across all sequences. This powerful combination helps detect outliers with precision and efficiency [79].

A SeqVAE-CNN model to carry out unsupervised deep learning for anomaly detection. This model takes inspiration from Variational Autoencoders (VAEs) and Convolutional Neural Networks (CNNs), creating a Seq2Seq structure that can capture both temporal relationships and spatial features in multivariate time-series data. The experimental results of their model on 8 datasets from different domains suggest it has a higher performance for anomaly detection; indeed, the highest AUROC and F1 scores have been observed when using our model [80].

Researchers propose a hybrid model of VAE-LSTM for unsupervised anomaly detection in time series. This model combines the features extracted from the VAE module [81], which capture local patterns for short windows, with the Long Short-Term Memory (LSTM) module, which captures long-term correlations in the time series. Additionally, Electrical power grids are vulnerable to cyber-attacks, existing attack detection methods are limited so to tackle Graph Convolutional Long Short-Term Memory (GC-LSTM) with a deep convolution network has been proposed to further improve time series classification and analysis with respect to anomaly detection and attack graph models [82].

A new hybrid anomaly detector that merges two detection approaches i.e., Key Performance Indicators (KPIs) that are used in physics and Unsupervised Variational Autoencoder (VAE), thereby improving accuracy and decreasing the possibility of overlooking defective elements in safety-critical scenarios. Performance is discussed in comparison to different VAE architectures like long short-term memory (LSTM-VAE) and bidirectional LSTM (BiLSTM-VAE). Additionally, the efficient choice of hyperparameters in these structures can be optimized with the help of a genetic algorithm as presented in reference [83].

Due to the many advantages of conventional anomaly detection in time series models, further innovations in this area have the potential to yield beneficial results. In fact, tackling the complexities associated with real-world time series requires advanced solutions, such as hybridization of hybrid classes. Research shows that this technique can provide great improvements in terms of forecasting accuracy and has been gaining much attention recently.

6. Forecasting and anomaly

Time Series Forecasting has always been a useful tool to detect trends, patterns of any data. It's about predicting the next time stamps using previous or existing trends. Anomaly Detection and Time series forecasting have been interlinked several times by researchers in this field. Several Machine Learning algorithms have been implemented, sometimes merged with each other to derive another novel strategy to predict whether the next time stamp is normal or abnormal.

The power of forecasting lies in its potential to revolutionize healthcare. The goal? To empower medical professionals to take proactive and timely action, reducing patient transfers and hospital stay lengths, ultimately leading to improved survival rates. But the accuracy of predictions relies heavily on expertly combining machine learning algorithms like autoencoders and extreme gradient boosting (XGBoost) [84].

Autoencoders excel in feature extraction [85]. They are uniquely adept at unsupervised anomaly detection when labeled data is scarce or nonexistent [86]. They're trained via reconstruction error, only triggering an alert if said error exceeds a predetermined threshold - prompting a swift remedial response. As for XGBoost, this decision tree-based ensemble principle takes physiological variables from time ti as input and outputs variables from the next temporal unit; ti+1 [84].

All told, tapping into modern technology's full potential could allow for massive improvements in healthcare outcomes - starting with careful utilization of solutions like autoencoders or XGBoost models.

Recent research has sought to compare the performance of supervised and unsupervised algorithms on physiological data. Heart rate data, due to its ubiquity and non-invasiveness, is ideal for predicting anomalies. Five algorithms were evaluated for detecting anomalies in heart rate -- two unsupervised techniques and three supervised methods. The models were tested on real heart rate data and findings demonstrated that both local outlier factor and random forests algorithms were effective in detecting abnormalities in this type of data. Additionally, results showed that simulated data can lead algorithms to a similar level as real labeled information when not available, enabling rapid initial deployment without prior knowledge [8].

DeepAnT is a deep learning-based anomaly detection approach for streaming and non-streaming time series data. It can detect a broad range of anomalies, from point anomalies to contextual and discords. Instead of learning about anomalies, DeepAnT uses unlabeled data to determine normal time series. The two key components of DeepAnT are its time series predictor (which uses CNN and takes context into account) and its anomaly detector module, which identifies whether an upcoming time stamp is normal or anomalous.

DeepAnT stands out against the competition by only needing a relatively small data set to generate a model. It utilizes parameter sharing of a convolutional neural network (CNN) which allows for good generalization capabilities. Unsupervised anomaly detection in DeepAnT removes the need for labeling, making it directly applicable to real-world scenarios with large streams of complex data from heterogeneous sensors. Neural networks are popular as they enable automatic feature discovery without having any prior domain knowledge; this capability is what makes them such excellent candidates for time series anomaly detection. DeepAnT optimizes through leveraging a CNN and raw data, making it more robust to variations than many other neural networks and statistical models on the market [87]. Using a data-driven approach can be beneficial in many contexts, especially when there is access to an abundance of untagged data. However, the data quality has a great impact factor on its accuracy; if too much of the dataset is contaminated (5% or more), then it could potentially lead to wrong inferences upon deployment. Additionally, selecting the right network architecture and hyperparameters are often difficult tasks. Nevertheless, new automated techniques have been developed that may assist in optimizing these settings instead of opting for human expertise [88]. Last but not least, one major drawback is the susceptibility to adversarial examples [89] which could restrict its usage in safety-critical system models. Luckily though, research into understanding and defending against such cases has increased progressively over time with some successful results achieved.

Light curve prediction and anomaly detection using LSTM neural networks is an important research area for time domain astronomy. A series of processing was done on star images collected from the National Astronomical Observatories of China using GWAC's mini-GWAC system, resulting in light luminance data over a period of time. Researchers explored a model of LSTM neural network to accurately predict light curves, with an optimal structure obtained through model training and validation; meanwhile, an anomaly detection mechanism based on prediction error was implemented. Results showed that this method has great potential when tested on real light curve data [90]. More historical data and certain well-known astronomical principles are needed to further improve upon this method.

Motorsports have limited access to sensors during competitions, limiting predictive capabilities and providing an edge for competitors. The proposed variational autoencoder-based selective prediction (VASP) framework addresses this challenge by combining the tasks of anomaly detection and time series prediction in one powerful approach. VASP consists of a variational autoencoder (VAE), an anomaly detector, and LSTM predictors which can all work together to help produce more robust predictions. Even if anomalies occur in the input signals, VASPs accuracy is not significantly impacted like that of other deep learning approaches such as long short-term memory (LSTM) neural networks. Try out VASP today to take your predictive insights to the next level with more effective technique [91].

7. Anomaly detection using AI

Time series data bring their own set of challenges when model analysis is applied, like notions of time and uncertainty, and the presence of drift. Typically, the time series window is broken down into two pieces with either sliding endpoints or landmark endpoints. In this paper, they categorize anomalies and outliers as the same, as presented in reference [92]. Detecting these outliers has been and remains an area of exploration for researchers and practitioners alike. Time series data is one of the most useful modalities available for a variety of applications. Upon analyzing this type of data, it becomes clear that outlier detection plays a key role. Companies such as Microsoft [93] have even created outlier detection services to monitor business data with triggers to alert them when outliers are present.

As stated in reference [94], AI assurance is an important process that must be incorporated throughout the engineering lifecycle of an AI system. This process should ensure the system is dependable and its outcomes valid, trustworthy, and ethical. Moreover, it should also be data-driven, explainable to all users, unbiased in its learning processes, and fair for all involved.

One of the recent hot algorithm in AI for Anomaly detection in time series is GAN proposed by Goodfellow [50], have become some of the most discussed topics in deep learning. The use of a generator in GANs helps to generate expected normal behavior, while a discriminator can distinguish between "normal" and "abnormal" behaviors. GAN technology has led to exciting new developments in deep learning. Generative adversarial networks (GAN) are an innovative form of AI that offer a powerful solution to the generative modeling problem. GAN is composed of two models - a generator used to create normal behavior and a discriminator used to distinguish between normal and abnormal behaviors. When dealing with imbalanced industrial time series data, GAN can be applied to derive an anomaly detection architecture that outperforms classic algorithms and other deep learning models such as big-GAN, ANOGAN and DBN [95]. The attached article further elaborates on the inner workings of GANs and their core design considerations. Additionally, drawing from research conducted by Li et al. [96], this architecture can feature a dynamic threshold generated by the discriminator which serves as a predictive warning for system failures or anomalies. GAN based approach is used to diagnose faults by generating much higher anomaly scores when a fault sample is fed into the trained model [95].

Moving on to next one GTAD, researchers have developed a new anomaly detection algorithm for multivariate time series, called Graph Attention Network and Temporal Convolutional Network for Multivariate Time Series Anomaly Detection (GTAD). This algorithm takes into account the correlation and temporal dependencies that many other existing algorithms fail to address. GTAD promises to provide better results when it comes to spotting anomalies in complex data sets. GTAD is an unsupervised approach powered by graph attention networks and temporal convolutional networks [97].

TadGAN is a breakthrough in unsupervised anomaly detection that makes use of Generative Adversarial Networks (GANs). At the core of the system are Long Short-Term Memory (LSTM) Recurrent Neural Networks, which provide an excellent base model for creating Generators and Critics. TadGAN is unique in its ability to capture temporal correlations with cycle consistency loss for more accurate time-series data reconstruction [98].

Future concerns: Combining information between different dimensions of multivariate time series is a key focus of future work [95] in AI algorithms. When it comes to GAN-based anomaly detection models, there can be difficulties in determining the right sliding window length and maintaining stability during training. Further research is needed in order to more effectively train GANs [50].

7.1 AI based toolkits for automated anomaly detectors

TODS: TODS is a comprehensive automated Time Series Outlier Detection System with a modular design that enables easy construction of pipelines. It includes a range of primitives for data processing, time series analysis, feature analysis, detection algorithms and reinforcement methods. This makes TODS suitable for both research and industrial applications [99, 100].

ANOVIZ: ANOVIZ is an innovative anomaly detection solution for multivariate time series. It provides you with accurate detections, as well as easy-to-use visualizations and user interfaces to promote better explanation and assessment of the quality of those detections [101].

AnomalyKiTS: AnomalyKiTS is a system that allows end users to detect anomalies in time series data. It provides a range of algorithms, as well as an enrichment module to label identified anomalies. AnomalyKiTS offers four categories of model building capabilities, enabling users to select the best option for their needs [102].

TranAD: TranAD is an advanced model designed to provide superior recognition and diagnosis results. Our proprietary focus score-based self-conditioning and adversarial training technology extract multi-modal features, while MAML ensures quick and efficient on-the-fly training with minimal data. With TranAD, you get the best of both worlds: powerful detection capabilities and superior performance. TranAD has been proven to outperform existing baseline methods [46]. There is a range of data sizes, formats, and anomalies to consider when deciding which anomaly detection toolkit to use.

Future Scrutinizes: To maintain the quality of pipeline discovery system, researchers are planning on adding more primitives in the future as well as improved integral searchers to ensure optimal performance. To incorporate predefined rules efficiently into pipelines, researchers should also aim to develop learning-based active learning techniques for our reinforcement module. Existing solutions may not be comprehensive enough for certain applications, such as scenarios where semi-supervised or prediction-based unsupervised anomaly methods are needed.

The current research focused on using machine learning techniques to detect anomalies before they arise in future forecasting, leveraging stacked and bidirectional LSTM. The analysis produced promising results as noted in reference [103], validating the use of such models for anomaly detection. The review of AI-based energy monitoring and anomaly detection commercial solutions for buildings [104] provides an overview of the available systems. Efficient predictive maintenance of equipment in various industries requires the detection of anomalies in time-varying multivariate data. Researchers presented MTV (Multivariate Time Series Visualization), a visual analytics system that helps to streamline collaboration between humans and AI, facilitating the most ideal workflow [105].

8. Conclusion

In conclusion, this chapter has provided an insight into the cutting-edge models for anomaly detection in time series, discussed their merits and pitfalls, and highlighted new areas of research that are being explored to solve unique problems posed

DOI: http://dx.doi.org/10.5772/intechopen.111886

by high-dimensional and complex data, high-volume data streams, and a need for real-time processing. These research areas have provided concrete examples of the applications of discussed models. Moreover, citations will help readers about how these models can be used in real world scenarios. We have also identified some of the current issues and suggested future directions for research concerning anomaly detection systems.

As the field of anomaly detection in time series continues to evolve and new challenges arise, it is crucial that researchers remain focused on developing innovative solutions that can effectively process high-dimensional and complex data in real-time. By better understanding the existing state-of-the-art models and the challenges that still need to be addressed, researchers can identify new opportunities for developing more effective anomaly detection systems.

We have not explored all the current algorithms, models, and new research areas. This chapter has just provided readers an overview of some current techniques and areas so that they can identify what is going on exactly in this field. Interested readers will definitely go out and do some more research about it and prepare themselves for further development within this growing field.

Acknowledgements

The authors acknowledge the editor of the book for his support throughout the writing process.

Conflict of interest

The authors declare no conflict of interest.

Notes/thanks/other declarations

The authors would like to thank the Editor of the book and the publisher for giving them a valuable opportunity to prepare a book chapter.

Author details

Farrukh Arslan*, Aqib Javaid, Muhammad Danish Zaheer Awan and Ebad-ur-Rehman
Department of Electrical Engineering, University of Engineering and Technology, Lahore, Pakistan

*Address all correspondence to: farrukh_arslan@uet.edu.pk

References

[1] Sen PC, Hajra M, Ghosh M. Supervised classification algorithms in machine learning: A survey and review. In: Advances in Intelligent Systems and Computing. Singapore: Springer Singapore; 2020. pp. 99-111

[2] Ezugwu AE, Ikotun AM, Oyelade OO, Abualigah L, Agushaka JO, Eke CI, et al. A comprehensive survey of clustering algorithms: State-of-the-art machine learning applications, taxonomy, challenges, and future research prospects. Engineering Applications of Artificial Intelligence. 2022;**110**(104743):104743

[3] Petropoulos F, Apiletti D, Assimakopoulos V, Babai MZ, Barrow DK, Ben Taieb S, et al. Forecasting: Theory and practice. International Journal of Forecasting. 2022;**38**(3):705-871

[4] Ratanamahatana CA, Lin J, Gunopulos D, Keogh E, Vlachos M, Das G. Mining time series data. In: Data Mining and Knowledge Discovery Handbook. Boston: Springer; 2009. pp. 1049-1077

[5] Fu T-C. A review on time series data mining. Engineering Applications of Artificial Intelligence. 2011;**24**(1):164-181

[6] Esling P, Agon C. Time-series data mining. ACM Computing Surveys. 2012;**45**(1):1-34

[7] Hilal W, Gadsden SA, Yawney J. Financial fraud: A review of anomaly detection techniques and recent advances. Expert Systems with Applications. 2022;**193**(116429):116429

[8] Šabić E, Keeley D, Henderson B, Nannemann S. Healthcare and anomaly detection: Using machine learning to predict anomalies in heart rate data. AI & Society. 2021;**36**(1):149-158

[9] Sharma B, Sharma L, Lal C. Anomaly detection techniques using deep learning in IoT: A survey. In: 2019 International Conference on Computational Intelligence and Knowledge Economy (ICCIKE). Dubai, United Arab Emirates: IEEE; 2019. pp. 146-149

[10] Gupta M, Gao J, Aggarwal CC, Han J. Outlier detection for temporal data: A survey. IEEE Transactions on Knowledge and Data Engineering. 2014;**26**(9):2250-2267

[11] Fox AJ. Outliers in time series. Journal of the Royal Statistical Society. 1972;**34**(3):350-363

[12] Bommert A, Sun X, Bischl B, Rahnenführer J, Lang M. Benchmark for filter methods for feature selection in high-dimensional classification data. Computational Statistics and Data Analysis. 2020;**143**(106839):106839

[13] Big data basics for digital marketers [Internet]. Gartner. [cited 28 April 2023]. Available from: https://www.gartner.com/en/marketing/insights/articles/big-data-basics-for-digital-marketers

[14] Blázquez-García A, Conde A, Mori U, Lozano JA. A review on outlier/anomaly detection in time series data. ACM Computing Surveys. 2022;**54**(3):1-33

[15] Ahmad S, Purdy S. Real-time anomaly detection for streaming analytics. arXiv [cs.AI]. 2016

[16] Barbariol T, Chiara FD, Marcato D, Susto GA. A review of tree-based approaches for anomaly

detection. In: Springer Series in Reliability Engineering. Cham: Springer International Publishing; 2022. pp. 149-185

[17] Nassif AB, Talib MA, Nasir Q, Dakalbab FM. Machine learning for anomaly detection: A systematic review. IEEE Access. 2021;**9**:78658-78700

[18] Schmidl S, Wenig P, Papenbrock T. Anomaly detection in time series: A comprehensive evaluation. Proceedings VLDB Endowment. 2022;**15**(9):1779-1797

[19] Kozitsin V, Katser I, Lakontsev D. Online forecasting and anomaly detection based on the ARIMA model. Applied Sciences (Basel). 2021;**11**(7):3194

[20] Tang H, Wang Q, Jiang G. Time series anomaly detection model based on multi-features. Computational Intelligence and Neuroscience. 2022;**2022**:2371549

[21] Xu H, Pang G, Wang Y, Wang Y. Deep isolation forest for anomaly detection. arXiv [cs.LG]. 2022

[22] Thill M, Konen W, Wang H, Bäck T. Temporal convolutional autoencoder for unsupervised anomaly detection in time series. Applied Soft Computing. 2021;**112**(107751):107751

[23] Fan J, Han F, Liu H. Challenges of big data analysis. National Science Review. 2014;**1**(2):293-314

[24] Toledano M, Cohen I, Ben-Simhon Y, Tadeski I. Real-time anomaly detection system for time series at scale. In: Anandakrishnan A, Kumar S, Statnikov A, Faruquie T, Xu D, editors. Proceedings of the KDD 2017: Workshop on Anomaly Detection in Finance. PMLR; 2018. pp. 56-65

[25] Mason A, Zhao Y, He H, Gompelman R, Mandava S. Online anomaly detection of time series at scale. In: 2019 International Conference on Cyber Situational Awareness, Data Analytics and Assessment (Cyber SA). Oxford, UK: IEEE; 2019. pp. 1-8

[26] Ranjan KG, Tripathy DS, Prusty BR, Jena D. An improved sliding window prediction-based outlier detection and correction for volatile time-series. International Journal of Numerical Modelling. 2021;**34**(1):e2816

[27] Zhai Y, Ong Y-S, Tsang IW. The emerging "big dimensionality". IEEE Computational Intelligence Magazine. 2014;**9**(3):14-26

[28] McNeil P, Shetty S, Guntu D, Barve G. SCREDENT: Scalable real-time anomalies detection and notification of targeted malware in mobile devices. Procedia Computer Science. 2016;**83**:1219-1225

[29] Lopez MA, Gonzalez Pastana Lobato A, Duarte OCMB, Pujolle G. An evaluation of a virtual network function for real-time threat detection using stream processing. In: 2018 Fourth International Conference on Mobile and Secure Services (MobiSecServ), Miami Beach, FL, USA; 2018. pp. 1-5. DOI: 10.1109/MOBISECSERV.2018.8311440

[30] Goncalves D, Bota J, Correia M. Big data analytics for detecting host misbehavior in large logs. In: 2015 IEEE Trustcom/BigDataSE/ISPA. Helsinki, Finland: IEEE; 2015

[31] Cui B, He S. Anomaly detection model based on Hadoop platform and weka interface. In: 2016 10th International Conference on Innovative Mobile and Internet Services in Ubiquitous Computing (IMIS). Fukuoka, Japan: IEEE; 2016. pp. 84-89

[32] Rettig L, Khayati M, Cudré-Mauroux P, Piorkówski M. Online Anomaly

Detection over Big Data Streams. In: Braschler M, Stadelmann T, Stockinger K, editors. Applied Data Science. Cham: Springer; 2019. DOI: 10.1007/978-3-030-11821-1_16

[33] Liu X, Nielsen PH. Regression-Based Online Anomaly Detection for Smart Grid Data. arXiv (Cornell University); 2016

[34] Xie S, Chen Z. Anomaly detection and redundancy elimination of big sensor data in Internet of things [Internet]. arXiv [cs.DC]. 2017

[35] Bhadani AK, Jothimani D. Big data: Challenges, opportunities, and realities. In: Effective Big Data Management and Opportunities for Implementation. IGI Global; 2016. pp. 1-24

[36] Hashem IAT, Yaqoob I, Anuar NB, Mokhtar S, Gani A, Ullah KS. The rise of "big data" on cloud computing: Review and open research issues. Information Systems. 2015;**47**:98-115

[37] Chandola V, Banerjee A, Kumar V. Anomaly detection. ACM Computing Surveys. 2009;**41**(3):1-58

[38] Erfani SM, Rajasegarar S, Karunasekera S, Leckie C. High-dimensional and large-scale anomaly detection using a linear one-class SVM with deep learning. Pattern Recognition. 2016;**58**:121-134

[39] Mirsky Y, Shabtai A, Shapira B, Elovici Y, Rokach L. Anomaly detection for smartphone data streams. Pervasive and Mobile Computing. 2017;**35**:83-107

[40] Sarker RA, Elsayed SM, Ray T. Differential evolution with dynamic parameters selection for optimization problems. IEEE Transactions on Evolutionary Computation. 2014;**18**(5):689-707

[41] Akoglu L, Tong H, Koutra D. Graph-based anomaly detection and description: A survey [Internet]. arXiv [cs.SI]. 2014

[42] Katal A, Wazid M, Goudar RH. Big data: Issues, challenges, tools and good practices. In: 2013 Sixth International Conference on Contemporary Computing (IC3). Noida, India: IEEE; 2013. pp. 404-409

[43] Shiravi H, Shiravi A, Ghorbani AA. A survey of visualization systems for network security. IEEE Transactions on Visualization and Computer Graphics. 2012;**18**(8):1313-1329

[44] Vaswani A, Shazeer NM, Parmar N, Uszkoreit J, Jones L, Gomez AN, et al. Attention Is all you Need. NIPS [Internet]; 2017 Available from: https://www.semanticscholar.org/paper/Attention-is-All-you-Need-Vaswani-Shazeer/204e3073870fae3d05bcbc2f6a8e263d9b72e776

[45] Xu J, Wu H, Wang J, Long M. Anomaly Transformer: Time series anomaly detection with Association Discrepancy [Internet]. arXiv [cs.LG]. 2021 [cited 28 April 2023]. Available from: http://arxiv.org/abs/2110.02642

[46] Tuli S, Casale G, Jennings NR. TranAD: Deep transformer networks for anomaly detection in multivariate time series data [Internet]. arXiv [cs.LG]. 2022 [cited 28 April 2023]. Available from: http://arxiv.org/abs/2201.07284

[47] Wang X, Pi D, Zhang X, Liu H, Guo C. Variational transformer-based anomaly detection approach for multivariate time series. Measurement (Lond) [Internet]. 2022;**191**(110791):110791 Available from: https://www.sciencedirect.com/science/article/pii/S0263224122000914

[48] Zhang H, Xia Y, Yan T, Liu G. Unsupervised anomaly detection

in multivariate time series through transformer-based variational autoencoder. In: 2021 33rd Chinese Control and Decision Conference (CCDC). Kunming, China: IEEE; 2021. pp. 281-286

[49] Kingma DP, Welling M. Auto-Encoding Variational Bayes [Internet]. arXiv [stat.ML]. 2013 [cited 28 April 2023]. Available from: http://arxiv.org/abs/1312.6114

[50] Goodfellow IJ, Pouget-Abadie J, Mirza M, Xu B, Warde-Farley D, Ozair S, et al. Generative Adversarial Networks [Internet]. arXiv [stat.ML]. 2014 [cited 28 April 2023]. Available from: http://arxiv.org/abs/1406.2661

[51] Chen Z, Chen D, Zhang X, Yuan Z, Cheng X. Learning graph structures with Transformer for multivariate time series anomaly detection in IoT [Internet]. arXiv [cs.LG]. 2021 [cited 28 April 2023]. Available from: http://arxiv.org/abs/2104.03466

[52] Wang C, Xing S, Gao R, Yan L, Xiong N, Wang R. Disentangled dynamic deviation transformer networks for multivariate time series anomaly detection. Sensors (Basel) [Internet]. 2023 [cited 28 April 2023];23(3):1104. Available from: https://www.mdpi.com/1424-8220/23/3/1104

[53] Wu B, Fang C, Yao Z, Tu Y, Chen Y. Decompose auto-transformer time series anomaly detection for network management. Electronics [Internet]. 2023;**12**(2):354. DOI: 10.3390/electronics12020354

[54] Zhou T, Ma Z, Wen Q, Wang X, Sun L, Jin R. FEDformer: Frequency Enhanced Decomposed Transformer for long-term series forecasting [Internet]. arXiv [cs.LG]. 2022 [cited 28 April 2023]. Available from: http://arxiv.org/abs/2201.12740

[55] Wu H, Xu J, Wang J, Long M. Autoformer:DecompositionTransformers with Auto-Correlation for long-term series forecasting [Internet]. arXiv [cs.LG]. 2021 [cited 28 April 2023]. pp. 22419-22430. Available from: https://proceedings.neurips.cc/paper/2021/hash/bcc0d400288793e8bdcd7c19a8ac0c2b-Abstract.html

[56] Zhang Y, Yan J. Crossformer: Transformer utilizing cross-dimension dependency for multivariate time series forecasting [Internet]. 2023 [cited 28 April 2023]. Available from: https://openreview.net/pdf?id=vSVLM2j9eie

[57] Nie Y, Nguyen NH, Sinthong P, Kalagnanam J. A time series is worth 64 words: Long-term forecasting with transformers [Internet]. arXiv [cs.LG]. 2022 [cited 28 April 2023]. Available from: http://arxiv.org/abs/2211.14730

[58] Cai L, Janowicz K, Mai G, Yan B, Zhu R. Traffic transformer: Capturing the continuity and periodicity of time series for traffic forecasting. Transactions in GIS[Internet]. 2020 [cited 28 April 2023];**24**(3):736-755 Available from: https://research-information.bris.ac.uk/en/publications/traffic-transformer-capturing-the-continuity-and-periodicity-of-t

[59] Xu M, Dai W, Liu C, Gao X, Lin W, Qi G-J, et al. Spatial-Temporal Transformer Networks for traffic flow forecasting [Internet]. arXiv [eess.SP]. 2020 [cited 28 April 2023]. Available from: https://paperswithcode.com/paper/spatial-temporal-transformer-networks-for

[60] Li L, Yao J, Wenliang L, He T, Xiao T, Yan J, et al. GRIN: Generative relation and intention network for multi-agent trajectory prediction. Advances in Neural Information Processing Systems [Internet]. 2021

[cited 28 April 2023];34:27107-27118. Available from: https://proceedings.neurips.cc/paper/2021/hash/e3670ce0c315396e4836d7024abcf3dd-Abstract.html

[61] Ding C, Sun S, Zhao J. MST-GAT: A multimodal spatial–temporal graph attention network for time series anomaly detection. Information Fusion [Internet]. 2023;**89**:527-536 Available from: https://www.sciencedirect.com/science/article/pii/S156625352200104X

[62] Devlin J, Chang M-W, Lee K, Toutanova K. BERT: Pre-training of deep bidirectional transformers for language understanding. In: Proceedings of the 2019 Conference of the North. Stroudsburg, PA, USA: Association for Computational Linguistics; 2019. pp. 4171-4186

[63] Brown TB, Mann B, Ryder N, Subbiah M, Kaplan J, Dhariwal P, et al. Language Models are Few-Shot Learners [Internet]. arXiv [cs.CL]. 2020 [cited 28 April 2023]. p. 1877-901. Available from: https://proceedings.neurips.cc/paper/2020/hash/1457c0d6bfcb4967418bfb8ac142f64a-Abstract.html

[64] Chen H, Wang Y, Guo T, Xu C, Deng Y, Liu Z, et al. Pre-trained image processing transformer. In: In: 2021 IEEE/CVF Conference on Computer Vision and Pattern Recognition (CVPR). IEEE; 2021. pp. 12299-12310

[65] Zerveas G, Jayaraman S, Patel D, Bhamidipaty A, Eickhoff C. A Transformer-based Framework for Multivariate Time Series Representation Learning. Proceedings of the 27th ACM SIGKDD Conference on Knowledge Discovery & Data Mining. 2021

[66] Yang C-HH, Tsai Y-Y, Chen P-Y. In: Meila M, Zhang T, editors. arXiv [cs.LG] Voice2Series: Reprogramming Acoustic Models for Time Series Classification [Internet]. 2021. pp. 11808-11819 Available from: https://proceedings.mlr.press/v139/yang21j.html

[67] Wu Z, Liu Z, Lin J, Lin Y, Han S. Lite Transformer with Long-Short Range Attention [Internet]. arXiv [cs.CL]. 2020 [cited 28 April 2023]. Available from: https://iclr.cc/virtual_2020/poster_ByeMPlHKPH.html

[68] Mehta S, Ghazvininejad M, Iyer S, Zettlemoyer L, Hajishirzi H. DeLighT: Deep and Light-weight Transformer [Internet]. openreview.net. 2023 [cited 28 April 2023]. Available from: https://openreview.net/forum?id=ujmgfuxSLrO

[69] Bapna A, Chen MX, Firat O, Cao Y, Wu Y. Training deeper neural machine translation models with transparent attention [Internet]. arXiv [cs.CL]. 2018 [cited 28 April 2023]. Available from: https://aclanthology.org/D18-1338.pdf

[70] Dehghani M, Gouws S, Vinyals O, Łukasz JU, Google K, Google B. UNIVERSAL TRANSFORMERS [Internet]. Arxiv.org. [cited 28 April 2023]. Available from: http://arxiv.org/abs/1807.03819v3

[71] Xin J, Tang R, Lee J, Yu Y, Lin J. DeeBERT: Dynamic early exiting for accelerating BERT inference. In: Proceedings of the 58th Annual Meeting of the Association for Computational Linguistics. Stroudsburg, PA, USA: Association for Computational Linguistics; 2020. pp. 2246-2251

[72] Dai Z, Yang Z, Yang Y, Carbonell J, Le QV, Salakhutdinov R. Transformer-XL: Attentive language models beyond a fixed-length context [Internet]. arXiv [cs.LG]. 2019 [cited 28 April 2023]. Available from: http://arxiv.org/abs/1901.02860

[73] So DR, Liang C, Le QV. The Evolved Transformer [Internet]. arXiv [cs.LG].

2019 [cited 28 April 2023]. Available from: http://arxiv.org/abs/1901.11117

[74] Chen M, Peng H, Fu J, Ling H. AutoFormer: Searching transformers for visual recognition. In: In: 2021 IEEE/CVF International Conference on Computer Vision (ICCV). IEEE; 2021. pp. 12270-12280

[75] Tkach V, Kudin A, Kebande VR, Baranovskyi O, Kudin I. Non-pattern-based anomaly detection in time-series. Electronics (Basel) [Internet]. 2023 [cited 28 April 2023];**12**(3):721 Available from: https://www.mdpi.com/2079-9292/12/3/721

[76] Zhou L, Zeng Q, Li B. Hybrid anomaly detection via multihead dynamic graph attention networks for multivariate time series. IEEE Access [Internet]. 2022;**10**:40967-40978 Available from: https://ieeexplore.ieee.org/abstract/document/9758699/

[77] Nguyen HD, Tran KP, Thomassey S, Hamad M. Forecasting and anomaly detection approaches using LSTM and LSTM autoencoder techniques with the applications in supply chain management. International Journal of Information Management [Internet]. 2021;**57**(102282):102282 Available from: https://www.sciencedirect.com/science/article/pii/S026840122031481X

[78] Mathonsi T, van Zyl TL. Statistics and deep learning-based hybrid model for interpretable anomaly detection [Internet]. arXiv [cs.LG]. 2022 [cited 28 April 2023]. Available from: http://arxiv.org/abs/2202.12720

[79] Nizam H, Zafar S, Lv Z, Wang F, Hu X. Real-time deep anomaly detection framework for multivariate time-series data in industrial IoT. IEEE Sensors Journal [Internet]. 2022;**22**(23):22836-22849 Available from: https://ieeexplore.ieee.org/abstract/document/9915308/

[80] Choi T, Lee D, Jung Y, Choi H-J. Multivariate time-series anomaly detection using SeqVAE-CNN hybrid model. In: 2022 International Conference on Information Networking (ICOIN). Jeju-si, Korea: IEEE; 2022. pp. 250-253

[81] Lin S, Clark R, Birke R, Schonborn S, Trigoni N, Roberts S. Anomaly detection for time series using VAE-LSTM hybrid model. In: ICASSP 2020-2020 IEEE International Conference on Acoustics, Speech and Signal Processing (ICASSP). Barcelona, Spain: IEEE; 2020. pp. 4322-4326

[82] Presekal A, Stefanov A, Rajkumar VS, Palensky P. Attack graph model for cyber-physical power systems using hybrid deep learning. IEEE Transactions on Smart Grid [Internet]. 2023:1-1 Available from: https://ieeexplore.ieee.org/abstract/document/10017381/

[83] Terbuch A, O'Leary P, Khalili-Motlagh-Kasmaei N, Auer P, Zohrer A, Winter V. Detecting anomalous multivariate time-series via hybrid machine learning. IEEE Transactions on Instrumentation and Measurement [Internet]. 2023;**72**:1-11 Available from: https://ieeexplore.ieee.org/abstract/document/10015855/

[84] Boloka T, Crafford G, Mokuwe W, Van Eden B. Anomaly detection monitoring system for healthcare. In: 2021 Southern African Universities Power Engineering Conference/Robotics and Mechatronics/Pattern Recognition Association of South Africa (SAUPEC/RobMech/PRASA). Potchefstroom, South Africa: IEEE; 2021. pp. 1-6

[85] Luo A, Yang F, Li X, Nie D, Jiao Z, Zhou S, et al. Hybrid graph neural networks for crowd counting. Proceedings of the AAAI Conference on Artificial Intelligence [Internet]. 2020 [cited 28 April 2023];**34**(07):11693-11700 Available from: https://ojs.aaai.org/index.php/AAAI/article/view/6839

[86] Goldstein M, Uchida S. A comparative evaluation of unsupervised anomaly detection algorithms for multivariate data. PLoS One [Internet]. 2016;**11**(4):e0152173. DOI: 10.1371/journal.pone.0152173

[87] Karadayi Y, Aydin MN, Ogrenci AS. Unsupervised anomaly detection in multivariate spatio-temporal data using deep learning: Early detection of COVID-19 outbreak in Italy. IEEE Access [Internet]. 2020;**8**:164155-164177 Available from: https://ieeexplore.ieee.org/abstract/document/9187620/

[88] Zoph B, Le QV. Neural architecture search with reinforcement learning [Internet]. arXiv [cs.LG]. 2016 [cited 28 April 2023]. Available from: http://arxiv.org/abs/1611.01578

[89] Kurakin A, Goodfellow I, Bengio S. Adversarial machine learning at scale [Internet]. arXiv [cs.CV]. 2016 [cited 28 April 2023]. Available from: http://arxiv.org/abs/1611.01236

[90] Zhang R, Zou Q. Time series prediction and anomaly detection of light curve using LSTM neural network. Journal of Physics: Conference Series. 2018;**1061**:012012

[91] von Schleinitz J, Graf M, Trutschnig W, Schröder A. VASP: An autoencoder-based approach for multivariate anomaly detection and robust time series prediction with application in motorsport. Engineering Applications of Artificial Intelligence [Internet]. 2021;**104**(104354):104354 Available from: https://www.sciencedirect.com/science/article/pii/S0952197621002025

[92] Haris M, Sharif U, Gupta K, Mohammed A, Jiwani N. Anomaly detection in time series using deep learning [Internet]. Ijeast.com. [cited 28 April 2023]. Available from: https://www.ijeast.com/papers/296-305%20Tesma0706.pdf

[93] Ren H, Xu B, Wang Y, Yi C, Huang C, Kou X, et al. Time-series anomaly detection service at Microsoft. In: Proceedings of the 25th ACM SIGKDD International Conference on Knowledge Discovery & Data Mining. New York, NY, USA: ACM; 2019

[94] Batarseh FA, Freeman L, Huang C-H. A survey on artificial intelligence assurance. Journal of Big Data [Internet]. 2021;**8**(1):60. DOI: 10.1186/s40537-021-00445-7

[95] Jiang W, Hong Y, Zhou B, He X, Cheng C. A GAN-based anomaly detection approach for imbalanced industrial time series. IEEE Access [Internet]. 2019;**7**:143608-143619 Available from: https://ieeexplore.ieee.org/abstract/document/8853246/

[96] Li D, Chen D, Jin B, Shi L, Goh J, Ng S-K. MAD-GAN: Multivariate anomaly detection for time series data with generative adversarial networks. In: Artificial Neural Networks and Machine Learning – ICANN 2019: Text and Time Series. Cham: Springer International Publishing; 2019. pp. 703-716

[97] He Y, Zhao J. Temporal convolutional networks for anomaly detection in time series. Journal of Physics: Conference Series. 2019;**1213**:042050

[98] Geiger A, Liu D, Alnegheimish S, Cuesta-Infante A, Veeramachaneni K.

TadGAN: Time series anomaly detection using generative adversarial networks. In: 2020 IEEE International Conference on Big Data (Big Data). Atlanta, GA, USA: IEEE; 2020. pp. 33-43

[99] Lai K-H, Zha D, Wang G, Xu J, Zhao Y, Kumar D, et al. TODS: An automated time series outlier detection system. Proceedings of the AAAI Conference on Artificial Intelligence [Internet]. 2021 [cited 28 April 2023];**35**(18):16060-16062 Available from: https://ojs.aaai.org/index.php/AAAI/article/view/18012

[100] Milutinovic M, Schoenfeld B, Martinez-Garcia D, Ray S, Shah S, Yan D. On Evaluation of AutoML Systems [Internet]. Automl.org. [cited 28 April 2023]. Available from: https://www.automl.org/wp-content/uploads/2020/07/AutoML_2020_paper_59.pdf

[101] Trirat P, Nam Y, Kim T, Lee J-G. ANOVIZ: A visual inspection tool of anomalies in multivariate time series [Internet]. Github.io. [cited 28 April 2023]. Available from: https://itouchz.github.io/files/AnoViz_AAAI23.pdf

[102] Patel D, Ganapavarapu G, Jayaraman S, Lin S, Bhamidipaty A, Kalagnanam J. AnomalyKiTS: Anomaly detection toolkit for time series. Proceedings of the AAAI Conference on Artificial Intelligence [Internet]. 2022 [cited 28 April 2023];**36**(11):13209-13211 Available from: https://ojs.aaai.org/index.php/AAAI/article/view/21730

[103] Girish L, Rao SKN. Anomaly detection in cloud environment using artificial intelligence techniques. Computing [Internet]. 2023;**105**(3):675-688. DOI: 10.1007/s00607-021-00941-x

[104] Himeur Y, Ghanem K, Alsalemi A, Bensaali F, Amira A. Artificial intelligence based anomaly detection of energy consumption in buildings: A review, current trends and new perspectives. Applied Energy [Internet]. 2021;**287**(116601):116601 Available from: https://www.sciencedirect.com/science/article/pii/S0306261921001409

[105] Liu D, Alnegheimish S, Zytek A, Veeramachaneni K. MTV: Visual analytics for detecting, investigating, and annotating anomalies in multivariate time series. Proceedings of the ACM on Human-Computer Interaction [Internet]. 2022;**6**(CSCW1):1-30. DOI: 10.1145/3512950

Section 2

Anomaly Detection vs Intrusion Detection

Chapter 5

Anomaly Detection through Adaptive DASO Optimization Techniques

Surendra Bhosale, Achala Deshmukh, Bhushan Deore and Parag Bhosale

Abstract

An intrusion detection systems (IDS) detect and prevent network attacks. Due to the complicated network environment, the ID system merges a high number of samples into a small number of normal samples, resulting in inadequate samples to identify and train and a maximum false detection rate. External malicious attacks damage conventional IDS, which affects network activity. Adaptive Dolphin Atom Search Optimization overcomes this. Thus, the work aims to create an adaptive optimization-based network intrusion detection system that modifies the classifier for accurate prediction. The model selects feature and detects intrusions. Mutual information selects feature for further processing in the feature selection module. Deep RNNs detect intrusions. The novel Adaptive Dolphin Atom Search Optimization technique trains the deep RNN. Adaptive DASO combines the DASO algorithm with adaptive concepts. The DASO is the integration of the dolphin echolocation (DE) with the atom search optimization (ASO). Thus, the intrusions are detected using the adaptive DASO-based deep RNN. The developed adaptive DASO approach attains better detection performance based on several parameters such as specificity, accuracy, and sensitivity.

Keywords: intrusion detection system (IDS), recurrent neural network (RNN), machine learning, anomaly detection, optimization algorithm

1. Introduction

Anomaly detection identifies the mechanism to study abnormal behavior notion in the data, events or experimental observations. Role of AI is very important the wide area of applications, including statistical analysis, video surveillance, computer vision, medical image analysis, neuroscience studies, financial fraud detection, law enforcement, and cyber securities.

Data analytics based on different ML novel algorithms opens a new field of research of complex data processing solutions, which can handle huge data set. Today, statisticians, programmers, engineers of multidiscipline, and medicos are brought on a common platform to have concrete and fast solution.

Real-time monitoring and dynamic data processing and finding the outliers in the dataset and finding the fact that do not conform to the normal behavior in a dataset. Cybercrimes are increasing day by day and shortcomings of the security protocols, and algorithms are exposed and worldwide internet-based businesses are affected [1].

Anomaly detection can be either supervised or unsupervised, depending on whether the dataset contains labeled or unlabeled data points. Anomaly detection system first trains machine learning algorithm (considering a dataset) to learn the normal patterns behavior, further it can identify the data points that deviate significantly from the normal behavior.

The role of AI in anomaly detection is, therefore, crucial, and new avenues of research in the field of data analytics can be possible and the demand for accurate and efficient anomaly detection algorithm will increase substantially [2].

Anomaly identification is important for cleaning up data, making sure it is secure, and building strong AI systems. This talk will talk about recent work in our group on (a) benchmarking existing algorithms, (b) building a theoretical understanding of how they work, (c) explaining anomaly "alarms" to a data analyst, and (d) re-ranking potential anomalies in response to analyst feedback. The talk will then talk about two applications: (a) identifying and diagnosing sensor failures in weather networks and (b) open category detection in supervised learning.

Anomaly detection is the process of finding events or trends that do not fit with what would normally happen. This is important for things as if predicted maintenance, but it can be hard to do just through inspection. Anomaly detection methods that use machine learning and deep learning (AI) can find things in time series or image data that would be hard to find any other way. Find out how and why to use anomaly detection methods to find strange things in sensor data from hardware.

2. Major research issues and areas in anomaly detection

In spite of several types of research carried out for anomaly detection, network security is encountered with several challenges some of them are as follows [3]: the system developed for anomaly detection in disabled many security features and automatic backup settings, erases stored data, and opens associations to get commands from a remote PC.

The multivariate statistical intrusion detection schemes in have difficulties in estimating distributions for high-dimensional data. The problems rely on the fact that it does not identify undesirable behavior, and thus the FAR can be high. The problems in include the reliance on a well-defined security policy, which may be absent, and its inability to detect intrusions that have not yet been made to be known to the IDS [4]. Some of the common challenges encountered by the network IDS are listed below:

- There are many types of intrusions; the detection techniques available cannot accurately detect these varying attacks.

- Intrusion detection is very challenging as most of the data are encrypted in the network; it is difficult to detect the attacks from an encrypted traffic.

- Dedicated hardware components are required for network intrusion detection, and these components are very expensive.

- Some IDS supports only in the recognition of network attacks, and the other system attack is not detected.
- Several intrusion detection mechanisms adopted find it difficult to detect the intrusions from the high-speed network.
- Insider attack is the most serious problem encountered by several IDS.
- It is very difficult to approximate when and what actions are going to occur in a system.

3. Recent novel contributions in anomaly detection area

Following are the broad area in which there is research work done in the last decade as developed based on "auto encoder-based techniques," "novel machine learning-based techniques," and "hybrid techniques in IDS." There are following four broad areas *viz* auto encoder-based techniques, supervised and unsupervised machine learning algorithm, hybrid techniques in IDS, and Deep learning-based approaches.

3.1 Auto encoder-based techniques

The auto encoder learning-based techniques utilized in intrusion detection system are demonstrated in this section.

Shone et al. [1] introduced a novel deep learning model for enabling NIDS operation within modern networks. This method was developed by grouping both shallow and deep learning, and it had the ability to correctly analyze broad-array of network traffic. Most principally, the influence of this nonsymmetric deep auto encoder (NDAE) and the accurateness, as well as speed of random forest algorithm were combined. Further, this method was evaluated practically and attaining promising outcomes. This method provided elevated stages of precision, recall, and accuracy and requires less time for training.

Mighan et al. [5] established a novel scalable IDS based on deep learning. Network-based IDS was mainly focused on this paper. This method used one of the best processing tools Apache Spark for quick identification malicious traffic and for big data. Furthermore, this system utilized a network of stack auto encoder (SAE), subsequently an SVM classifier. For an underlying extraction of feature, the SAE was used. This methodology had four stages, known as, data preprocessing stage, decision-making stage, latent feature extraction stage, and attack classification stage. The stage of data preprocessing was the responsible for the preprocessing of data for making it ready for the extraction of feature.

Andresini et al. [6] established an auto encoder-based deep metric learning for network intrusion detection and the invented intrusion detection strategy evaluates the flow-based characteristics of the data of network traffic. The model of intrusion detection was learned by influencing a deep metric learning approach, which initially united the triplet networks, as well as auto encoders. Two distinct auto encoders were trained in the training stage on historical normal network flows, as well as attacks correspondingly. After that, a triplet network was trained for learning the inserting of the network flows' feature vector demonstration.

3.2 Novel machine learning-based techniques

The novel machine learning-based techniques utilized in intrusion detection are demonstrated in this section.

Kaja et al. [7] discovered a new two-stage intelligent IDS for the detection and protection of systems from such malicious attacks. Four preprocessing steps were performed in this method. Initially, the variance of values of feature was calculated for calculating the increase between features present in the dataset. The correlated features are estimated and eliminated in the second step of preprocessing in order to avoid overfitting. The third step utilized least square regression error (LSQE) to maximize the reduction of dimensionality and to minimize similarities of feature. At last, for analyzing relationships of feature, the maximal information compression index (MICI) was utilized.

Jin et al. [8] implemented a Bayes system based on light gradient boosting machine (light GBM) and parallel intrusion detection mechanism. The developed IDS was called as Swift IDS, which had the ability of both investigation of enormous data of traffic in speedy networks well-timed along with maintenance of reasonable performance of detection. The abovementioned goals were achieved by Swift IDS through two approaches. Light GBM was adopted in one approach as the algorithm of detecting intrusion for the management of enormous traffic data.

Sarker et al. [9] presented a machine learning-based security model namely, intrusion detection tree (IntruDTree). In this approach, initially, the security features ranking was considered in accordance with their significance in modeling. Afterward, a comprehensive model of intrusion detection based on a tree was constructed on the basis of the chosen significant features. After completing the construction of the entire tree by means of the integrating security data, the test data was utilized for authenticate the model.

Injadat et al. [10] developed a multistage optimized machine-learning framework for network intrusion detection. The effect of oversampling methods on the training model size of the models was studied initially and the least reasonable training size for successful intrusion identification was determined. This article suggests a multi stage enhanced machine learning-based NIDS structure, which decreases computational difficulty while keeping up with its performance of recognition. The stage of data preprocessing includes the process of normalization of data utilizing the Z-score strategy in addition to minority class oversampling by the usage of synthetic minority oversampling technique (SMOTE) algorithm.

Bertoli et al. [11] illustrated an end-to-end framework for machine learning-based NIDS. The AB-TRAP architecture was described in this paper, which enabled the application of updated network traffic, as well as think about the operational worries for enabling the entire employment of the resolution. The utilized AB-TRAP was a framework had five steps, comprising of the creation of the attack dataset, implementation of the models, the bonafide dataset, training of machine learning models, and the evaluation of performance of the recognized model following employment.

3.3 Hybrid techniques in IDS

The hybrid learning-based techniques exploited in intrusion detection are demonstrated in this section.

Jiang et al. [12] established a network intrusion detection algorithm combined hybrid sampling with deep hierarchical network for the improvement of detection accuracy. Two parts were included in the hybrid sampling. At first, for the elimination of noise samples in the mainstream class, the one side selection (OSS) algorithm was utilized. Next, the SMOTE was employed for creating the minority class samples in order to lighten the imbalance of samples of minority class. Therefore, for the classification, the imbalanced data was converted into balanced data. Deep hierarchical network was constructed for the difficulty of data features, which train the CNN in addition to bi-directional long short-term memory (Bi-LSTM) when learning the temporal and spatial feature of network traffic data.

Cavusoglu [13] implemented a new hybrid approach for intrusion detection using machine learning methods. The developed IDS used a mixture of various feature selection, as well as machine learning-based methods for offering high performance intrusion discovery in several types of attacks. The first step of this technique was data preprocessing, then the dataset's size was decreased by utilizing various feature selection algorithms. For feature selection, two novel methods were introduced. By the determination of suitable machine learning algorithms according to type of attack, the layered architecture was generated. This approach had low false positive rates and high accuracy in every form of attacks.

3.4 Deep learning-based approaches

The techniques on the basis of deep learning utilized in recognition of intrusion are demonstrated in this division. This section is again classified into three as follows:

3.4.1 CNN-based techniques

The CNN learning-based methods employed in intrusion detection are demonstrated in this division.

Tang et al. [14] developed a deep learning method for network intrusion detection in software-defined networking (SDN). In the established framework, the module of NIDS was established in the controller. The entire open flow switches were monitored by the SDN controller and called all network information when required; hence, the benefit of this global network was taken by the NIDS module for the detection of intrusion. After a fixed time window, a request message was sent to the entire open flow switches from the controller for requesting the network information.

Wu et al. [15] introduced an original intrusion detection model for a huge network using CNN. So as to involuntarily choose traffic features from raw dataset the CNN was utilized and the cost function weight coefficient of each category was set on the basis of its numbers for solving the problem of unprovoked dataset. Standard datasets were utilized by this approach for assessing the performance of the developed CNN model. The raw format of traffic vector was altered into image format in order to condense the cost for calculation. This method reduced the false alarm rate and improved the calculation cost and accuracy.

3.4.2 DNN-based method

The DNN learning-based methods exploited in intrusion detection are demonstrated in this division.

Vinayakumar et al. [16] presented a deep-learning approach for and intelligent IDS. This method utilized deep learning network (DNN) for designing an effective and flexible IDS for detecting and classifying unexpected and unknown cyber-attacks. The summary and high-dimensional feature demonstration of the IDS information are learned by sending them into several hidden layers. Moreover, this approach took up a multilayer perceptron (MLP) model that was a form of feed-forward neural (FFN) network consisting three or more than three layers with one output layer, one or more hidden layer, and one input layer wherein every layer had a lot of units or neurons in mathematical notation.

Gao et al. [17] explored an adaptive ensemble machine learning model for intrusion detection in which the advantages of every algorithm for various form of data detection was integrated, and optimal results are achieved through ensemble learning. The benefit of ensemble learning was merging the guesses of various fundamental estimators to enhance the robustness and generalize ability over a distinct estimator. A few widespread algorithms are utilized in this approach such as DNN, decision tree, and random forest to train this model. Also, the adaptive voting and multi-tree algorithm are developed in order to enhance the consequence of intrusion detection. It was found from the comparison with various existing methods; this method was superior to many former research outcomes and had good application prospects.

3.4.3 Other techniques

The other deep learning-based techniques utilized in intrusion detection are demonstrated in this section.

Otoum et al. [18] developed deep learning-based intrusion detection in the supervising of critical infrastructures through sensor networks. The main intention of this research is to examine the prospective of deep learning as a substitute for IDS based on robust machine learning. A restricted Boltzmann-based Clustered IDS (RBC-IDS) model was presented for a deep learning solution for detecting intrusion in critical network applications based on wireless sensor network.

Yang et al. [19] introduced a joined wireless network intrusion detection model in view of deep learning. The deep belief network (DBN) was involved in this approach as the layer of feature extraction and support vector machine (SVM) characterization layer. DBN layer, the error backpropagation algorithm, and the contrast divergence algorithm were utilized to decrease aspects of information and extort features. It assisted SVM for enhancing the capability to categorize high-dimensional information. Contrasted with the preceding forward proliferation, this approach changed the credence of the multi-restricted Boltzmann machine (RBM) with the back-propagation algorithm. The recognition model was prepared and laid out with the SVM to keep on enhancing the interruption. The classification performance of DBN was efficiently progressed by this approach. Thus, the precision rate, recall rate, and accuracy rate of this approach were superior to other methods.

Wu and Guo [20] introduced a hierarchical deep neural network for network intrusion detection namely, LuNet, which was made up of numerous levels of merged recurrent convolution sub-nets. The input data at every level was learned by RNN and CNN nets. The granularity of learning turned into increasingly detailed, as progress of the learning from the first level to the last level. With an understanding, the synergy of both RNN and CNN was efficiently utilized for the of both temporal and spatial feature extractions. By means of an in-depth examination and conversation for the arrangement of LuNet, high learning efficiency was attained by this method.

Khan et al. [21] presented a novel two-stage deep learning (TSDL) model, in view of a stacked auto-encoder with a soft-max classifier, for effective network intrusion identification. Two decision stages are involved in this model. This model varies from preceding models as it comprised two stages of feature representation. Feature representation was learned by the primary stage for characterizing typical and unusual network traffic with a possibility score value.

Sohi et al. [22] presented a recurrent neural network-based IDS, namely RNNIDS to catch complex designs in attacks and generate like ones. Initially, by the application of a new method, the generation of mutants of a malware was demonstrated, and this was the first step of this approach. This approach depends on the truth that an unknown pattern could be learned and extorted by a RNN. On the basis of this truth, new and unseen sequences were generated.

Zeng et al. [23] invented a deep learning-based network encrypted traffic classification and intrusion detection framework for detecting intrusions. The developed architecture was named as Deep-Full-Range (DFR), and it had three deep-learning algorithms such as CNN, LSTM, and SAE. The CNN was used for learning features of the raw traffic from spatial range. The features were learned from the time-related aspect by the use of LSTM. The SAE was taken for extracting features from coding characteristics.

4. Proposed adaptive dolphin atom search optimization-based DRNN for network intrusion detection system

The main challenges in network security are the development of efficient and robust NIDS. Although the significant developments in NIDS technology, the majority of solutions are still functioning by less capable signature-dependent techniques as opposed to anomaly detection approaches. The recent situation reaches a point, whereby reliance on such techniques leads to unsuccessful and inaccurate analysis. These challenges are utilized to create a widely-accepted anomaly detection (AD) technique capable of overcoming limitations caused by the ongoing changes happening in modern networks. NIDS is composed of data gathering, attribute extraction, attribute selection, IDS, and report generation. Every component in IDS have own impacts and functions, which are not noticed. There are three major limitations exist in IDS, where the contribution of this ID system is related to these limitations. First challenge relies on the enormous quantity of network information, and this issue can be handled using developing technique, which evaluates the data in an efficient manner. Second challenge is granularity and depth observing required for boosting up efficiency. Third limitation relies on quantity of distinct protocols and enormous quantity of data communication through traditional networks that commence large levels of intricacy and complexity. This augments the complexity for evaluating an exact scope of potential implementation or zero-day attacks [24].

Machine learning (ML) techniques are enormously adapted for recognizing distinct categories of attacks, and ML technique assists the system supervisor acquire the respective measures for preserving intrusions. Nevertheless, majority of conventional ML techniques be owned by superficial learning, which cannot successfully evaluate the issue of enormous intrusion information [25]. These limitations arise in the features of real system in application background. The invention of multi-classification process diminished accuracy with effective development of dataset. Additionally, superficial learning is inappropriate to knowledge-based analysis and broadcasting necessities of elevated dimensional learning accompanied with enormous data. In contrast, deep learners have ability for extracting better illustration from review data for generating better learning schemes. Consequently, IDS has familiarized rapid improvement after diminishing into moderately slow period [26]. Though, majority of traditional ML techniques related to superficial learning and regularly emphasize selection and feature engineering [27]. The innovations of deep learning (DL) techniques employ a rapid improvement in the recent period, which gives an improvement for detecting the new IDS. Recurrent neural network (RNN) plays a significant function in enhancement of DL techniques in the domain of language processing, translation, image depiction, human behavior identification, and semantic realization [6]. Since, DL approaches contain potential for identifying better illustrations from the information for creating much better schemes and inspired by RNN [28].

The main aspire of research is the detection of intrusions exist in the network using DASO-based deep RNN. Initially, input image is fed into the feature selection using mutual information in which the relevant features are selected. Then, the selected features are forwarded to the ID module in which the process is done by deep RNN. The deep RNN is trained using adaptive DASO algorithm for predicting whether it is intruder or not.

Proposed model: The main contribution of the research is development of adaptive DASO-based deep RNN for intruder detection. The Adaptive DASO is utilized to train the DRNN for predicting whether the network is intruder or not.

4.1 Developed adaptive DASO-based DRNN for NIDS

NIDS is the efficient method for preserving the computer networks from malicious threats and attacks. Different ID methods are adapted to predict the behavior of malicious activities, but an accurate detection of intrusion exist in the network system offers a major challenge. To deal with this challenge, an effective optimization method, named adaptive DASO-based DRNN is developed for identifying intrusion behavior in the network. The developed optimization scheme completes ID approach using two stages such as feature selection and ID. Initially, input data is gathered from ID dataset, and then it is forwarded to the feature selection steps, where the relevant features are selected using mutual information. Once the suitable features are selected, the intrusions are detected using DRNN, where the weight of the classifier is trained using developed Adaptive DASO algorithm for predicting the malicious behavior. Adaptive DASO algorithm is designed by including adaptive concept with the integration of DE and ASO. **Figure 1** shows the schematic illustration of developed model.

4.2 Get the input data

Let us choose the database as F with x number of network intrusion data D, which is depicted as,

$$F = \left\{D_1, D_2, \dots, D_p, \dots D_x\right\} \tag{1}$$

where, D depicts the intrusion data, F indicates the database, and D_p demonstrates the intrusion data situated at p^{th} index. From the intrusion dataset, intrusion data of

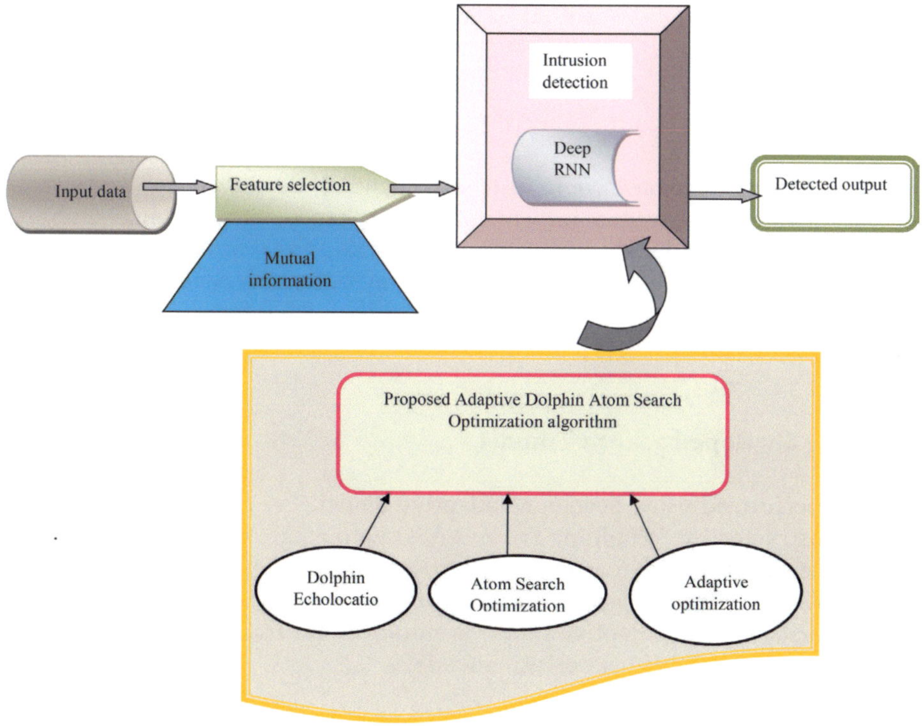

Figure 1.
Schematic illustration of developed model for ID.

input network D_p is considered and is permitted to feature selection module for performing ID process.

4.3 Selection of features through mutual information

The input data D_p is gathered from database and is fed to the feature selection stage, where the important features are successfully extracted to reduce the dimensionality of data. Here, the feature selection module is modeled using theory of mutual Information [29], which is employed to overcome the nuisance of dimensionality in prediction of malicious system. The motive of feature selection is to extract the relevant features suitable for identifying the behavior of intrusions. Mutual information establishes the relation among class label and features that are sampled simultaneously for predicting the relevant features. According to information theory [30], the mutual information among two constraints is nothing if and only if two constraints are statistically autonomous. Joint distribution of mutual information among two features S, and class label T is evaluated as,

$$L(S,T) = \sum_{s\varepsilon S} \sum_{t\varepsilon T} L(s,t) \log \frac{L(s,t)}{L(s).L(t)} \tag{2}$$

$L(S)$ and $L(T)$ depict borderline distributions of S and T formed through marginalization approach. Here, S shows the features, and T shows the labels of class. Finally, the features are chosen by the mutual information theory are represented as S and expressed by,

$$S = \{S_1, S_2, \dots S_m, \dots S_n\} \tag{3}$$

Here, S_n indicates the total count of features, and S_m depicts the mth feature. The output attained from the mutual information theory can be either normal or abnormal behavior, which is considered based on threshold value for choosing the features [31]. The relevant features selected using mutual information are denoted as S with the dimension of $[1 \times n]$. Furthermore, the features chosen from feature selection are fed to the input of deep RNN for performing ID process.

4.4 ID using developed adaptive model

NIDS is performed using developed adaptive model. The developed Adaptive DASO is constructed by combining DE and ASO with adaptive concept. The deep RNN classifier takes feature S as input obtained from feature selection module and initiates intrusion detection process with the hidden layers of neural network. Furthermore, the developed Adaptive DASO is employed for training the weights of classifier for achieving optimal performance [32].

4.4.1 Structure of DRNN

Deep RNN uses the information from the feature selection tool to do its work. It has three levels, including the input layer, the hidden layer, and the output layer. In neural network design, the input layer is at the top and the output layer is at the bottom. The hidden layer is in the middle. The output pattern of the last layer is fed into the first layer of the next layer, and so on. The repeating link is only made between levels that are hidden. Deep RNN classifier is better because it takes less time to learn the data. The system design of deep RNN is depicted in **Figure 2**.

The organization of DRNN classifier is formed by picking the input vector of ith layer at j^{th} time as $S^{(i,j)} = \left\{S_1^{(i,j)}, S_2^{(i,j)}, S_3^{(i,j)}, \dots S_h^{(i,j)}, \dots, S_n^{(i,j)}\right\}$ and output vector of ith layer at j^{th} time as $R^{(i,j)} = \left\{R_1^{(i,j)}, R_2^{(i,j)}, R_3^{(i,j)}, \dots R_h^{(i,j)}, \dots, R_n^{(i,j)}\right\}$, respectively. h represents arbitrary unit number of ith layer, and n species the total count of units of ith layer. Moreover, the arbitrary unit number, total number of units of $(i-1)^{th}$ layer is indicated as i and j, respectively However, the elements of the input vector are demonstrated as,

$$S_h^{i,j} = \sum_{z=1}^{q} r_{hz}^{i} R_z^{(i-1,j)} + \sum_{h'}^{n} u_{hh'}^{i} R_{h'}^{(i,j-1)} \tag{4}$$

where, r_{hz}^i and $u_{hh'}^i$ are the elements of $G^{(i)}$ and $g^{(i)}$. Arbitrary unit number of ith layer is represented as h'. The elements of the output vector of ith layer are expressed as,

$$R_h^{(i,j)} = \chi^{(i)}\left(S_h^{(i,j)}\right) \tag{5}$$

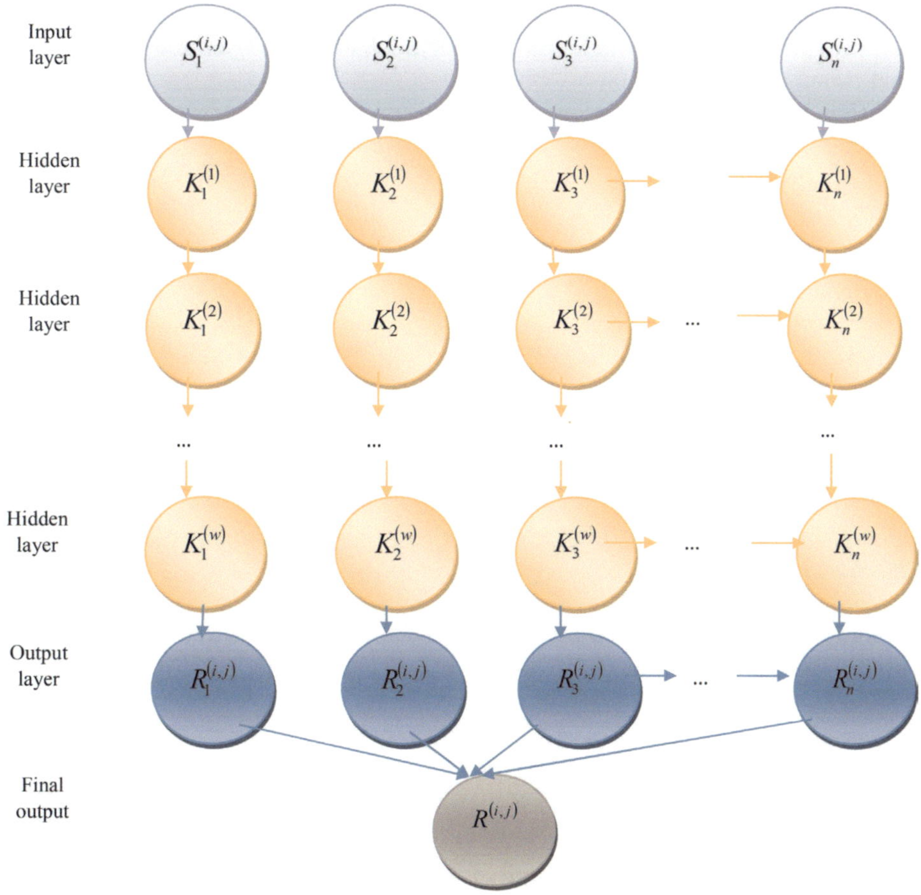

Figure 2.
System design of deep RNN classifier.

where, $\chi^{(i)}$ depicts the activation function. After assessing the output vector, the look becomes

$$R^{(i,j)} = \chi^{(i)}\left(G_i R^{(i-1,j)} + g^{(i)} R^{(i,j-1)}\right) \quad (6)$$

Here, $R^{(i,j)}$ indicates the output of classifier.

4.4.2 Training process of Deep RNN using adaptive DASO algorithm

The DRNN classifier is trained using developed Adaptive DASO. The developed Adaptive DASO is formed by including adaptive concept with the integration of DE and ASO. The ASO method is developed using the attraction and repulsion behavior of atoms. Every atom interacts with other atoms using attraction behavior and repulses the premature and over-concentrated atoms using repulsion properties. On the other hand, the DE method is introduced for enhancing the security of detection using the behavior of dolphins. DE method investigates the large space of candidate solutions, and it is performed till the global solution is achieved. The mixture of DE and ASO scheme, called as DASO technique, that offers best solution to solve optimization issues; however, this method consumes more computational time. Hence, in this research adaptive concept is included with DASO method for obtaining less computational time.

Solution encoding: The developed optimization is employed to estimate the optimal solution and reduced the error rate for NIDS-based on fitness measure. However, the implementation steps engaged in the developed adaptive model are summarized as below:

4.4.2.1 Population initialization

Let υ be the number of atoms and the position of dth atom is depicted as,

$$I_d = \left[I_d^1, \dots, I_d^a\right]; d = [1, \dots, l] \tag{7}$$

where, I_d^a denotes the ath position component of dth atom.

4.4.2.2 Fitness function

Fitness function is evaluated by estimating the variation of predicted output and classifier output, and the less error value is selected as the best solution, which is expressed as,

$$\sigma_d = \frac{1}{\mu}\left[\sum_{s=1}^{\mu} \varepsilon_s - R_s^{(i,j)}\right] \tag{8}$$

where, σ_d indicates the fitness value of dth atom, $R_s^{(i,j)}$ depicts the classifier output, and $\varepsilon_s$denotes the predicted output.

4.4.2.3 Mass computation

Atom mass is estimated using fitness function and mass of dth atom at f^{th} iteration is specified as,

$$M_d(f) = \frac{e_d(f)}{\sum_{d=1}^{l} e_f(f)} \tag{9}$$

where, $M_d(f)$ indicates the mass, and the term $e_d(f)$ is expressed as,

$$e_d(f) = \frac{\sigma_d - \sigma_{best}}{e^{\sigma_{worst} - \sigma_{best}}} \tag{10}$$

where, the terms σ_{best} and σ_{worst} specifies the best and worst value, and the expression is depicted as,

$$\sigma_{best} = \min_{d=1,\dots,l} \sigma_d \tag{11}$$

$$\sigma_{worst} = \max_{d=1,\dots,l} \sigma_d \tag{12}$$

4.4.2.4 Evaluate neighbor

The exploration of initial iteration is enhanced by selecting the N neighbors, which is based on the fitness value of interactions between atoms. The expression for N is depicted as,

$$N(f) = l - (l - 2)\sqrt{\frac{f}{d}} \tag{13}$$

4.4.2.5 Calculate the total force and constraint force

The summation of overall component that performed on the dth atoms from neighboring atoms is specified as total force, and the expression given by,

$$Q_d^a(f) = \sum_{s\varepsilon N_{best}} rand_s Q_{ds}^a(f) \tag{14}$$

where, $Q_d^a(f)$ indicates the force, and the term $rand_s$ specifies the random number and varies from 0 to 1, respectively. Every atom in the population space behaves as the best atom along with the constraint force of dth atom is expressed as,

$$\lambda_d^a(f) = H(f)\left(I_{best}^a(f) - I_d^a(f)\right) \tag{15}$$

where, $H(f)$ indicates the lagrangian multiplier.

4.4.2.6 Estimate the acceleration

The acceleration of dth atom at f^{th} time is calculated as,

$$A_d^a(f) = \frac{Q_d^a(f)}{M_d^a(f)} + \frac{\lambda_d^a(f)}{M_d^a(f)} \tag{16}$$

where, $Q_d^a(f)$ is the total force, $\lambda_d^a(f)$ is the constraint force, $M_d^a(f)$ indicates the mass, and $A_d^a(f)$ indicates acceleration of dth atom at f^{th} time.

4.4.2.7 Renew the velocity

The velocity of dth atom at $f + 1$ iteration is expressed as,

$$V_d^a(f + 1) = rand_d^a V_d^a(f) + A_d^a(f) \tag{17}$$

where, ${rand_d}^a$ indicates the random number, and $A_d^a(f)$ specifies the acceleration.

4.4.2.8 Update the atom location

The final updated equation of DASO algorithm is given as follows.

$$I_d(f+1) = \frac{\omega_{2d} M_d(f)}{\omega_{2d} M_d(f) - Ze^{\frac{-20f}{\alpha}}} \begin{bmatrix} I_d(f) + rand_d\, V_d(f) - \psi\left(1 - \frac{f-1}{\alpha}\right)^3 e^{\frac{-20f}{\alpha}} \sum\limits_{s \in N_{best}} \frac{rand_s\left[2 \times (c_{ds}(f))^{13} - (c_{ds})^7\right]}{M_d(f)} \\ \frac{(I_s(f) - T_d(f))}{\|I_d(f) I_s(f)\|_2} - Ze^{\frac{-20f}{\alpha}} \frac{I_d(f) + W_d(f) + \omega_{1d} J_d - \omega_{1d} I_d(f)}{\omega_{2d} M_d(f)} \end{bmatrix} \tag{18}$$

where, $M_d(f)$ specifies the mass of dth atom, $V_d(f)$ is the velocity, Z indicates the multiplier weight, ψ specifies the depth weight, α shows the maximum iteration, W_d signifies the search space dimension, J_d depicts the personal best solution, and ω_{1d} and ω_{2d} are the random number that lies between 0 to 1.

In equation, the term ψ is made adaptive for better performance of intrusion detection. The expression ψ is given by,

$$\psi = \psi_{\max} - \frac{f(\psi_{\max} - \psi_{\min})}{\alpha} \tag{19}$$

where, α signifies the depth weight, which is made adaptive, $\psi_{\max}$ and $\psi_{\max}$ depicts the predefined max, and min value of ψ and α signifies the highest iteration. Algorithm 1 states the pseudocode of the developed adaptive model.

4.4.2.9 Re-compute the fitness

Fitness value is predicted using objective function, which is mentioned in Eq. (8), where the fitness with optimal value is selected as optimal solution.

4.4.2.10 Termination

The abovementioned iteration is repeated until the stopping criteria are reached. The pseudocode of developed adaptive DASO-based deep RNN techniques is specified in Algorithm 1.

Algorithm 1. Pseudocode of the developed adaptive model.

Sl. no	**Pseudocode of the developed adaptive model**
	Input: I_d
	Output: $I_d^a(f+1)$
1	Initiate the set of atoms I and the velocity V
2	While termination criteria are not satisfied
3	Do
4	Evaluate σ
5	if $(\sigma_d < \sigma_{best})$then
6	$\sigma_{best} = \sigma_d$
7	$I_{best} = I_d$

DOI: http://dx.doi.org/10.5772/intechopen.112421

Sl. no	Pseudocode of the developed adaptive model
8	End if
9	Calculate $M_d(f)$
10	Determine N neighbors
11	Compute $Q_d^a(f)$ and $\lambda_d^a(f)$
12	Calculate $A_d^a(f)$
13	Update $V_d^a(f+1)$
14	Location update of atom $I_d^a(f+1)$using Eq. (18)
15	Introduce adaptive concept in place of $\psi = \psi_{\max} - \frac{f(\psi_{\max} - \psi_{\min})}{\alpha}$
16	End for
17	End while
18	Return I_{best}

By including the Adaptive concept with ASO and DE provides enhanced optimal result, and the computation time is also reduced. The performance of intrusion detection is also enhanced by including the adaptive concept within the hybrid optimization algorithm.

5. Results and discussion

The results of developed adaptive model are briefly discussed in this area in terms of sensitivity, accuracy, and specificity.

5.1 Experimental setup and dataset description

The developed adaptive model is executed in Pythontool using NSL-KDD dataset [33], and BoT-IoT dataset [34]. Dataset-1 includes multiple information for solving the optimization troubles such that this information is reasonable. The Dataset-2 comprises the source files with different formats such as CSV files, argus files, and pcap files. However, these files are partitioned based on the kind of attacks.

5.2 Evaluation parameters

The performance parameters utilized for the analysis of intrusion detection in the proposed adaptive model are sensitivity, accuracy, and specificity.

5.2.1 Sensitivity

The sensitivity is the proportion of true positive (TP) to the addition of TP and false negative (FN). The sensitivity is expressed as,

$$Sensitivity = \frac{P_T}{N_F + P_T} \tag{20}$$

5.2.2 Accuracy

The accuracy is the degree of proximity between predicted and original value. The accuracy is expressed as,

$$Accuracy = \frac{N_T + P_T}{P_F + N_F + P_T + N_T} \tag{21}$$

5.2.3 Specificity

The specificity is the proportion of true negative (TN) to the addition of false positive (FP) and true negative (TN). The specificity is termed as,

$$Specificity = \frac{N_T}{N_T + P_F} \tag{22}$$

where, P_T, P_F, N_T and N_F represented the true positive, false positive, true negative, and false negative, respectively.

5.3 Comparative methods

The performance of the developed method is analyzed by comparing developed method with the other state-of-the-art techniques, such as DBN [1], CNN [13], as well as DSAE [14], respectively.

5.4 Comparative analysis

This part talked about how the developed adaptive DASO-based DRNN with dataset-1 and dataset-2 were compared.

5.4.1 Analysis using dataset-1

Figure 3a shows how accuracy can be looked at by changing the training data. For 60% of training data, the accuracy of the newly created adaptive model is 0.8856, while the accuracy of the currently used methods, such as DBN, DSAE, CNN, and DASO-based DRNN, is 0.8290, 0.8224, 0.8056, and 0.860317, respectively. The performance of the adaptive DASO-based deep RNN was improved by 6.39354%, 7.1329%, 9.0376%, and 2.8613% when compared to state-of-the-art methods such as DBN, DSAE, CNN, and DASO-based deep RNN.

Figure 3b shows how the sensitivity and training data were looked at. The created adaptive model has a sensitivity of 0.9849, while the training data is 70%. With existing methods, such as DBN, DSAE, CNN, and DASO-based deep RNN, the sensitivity values are 0.9362, 0.9230, 0.89, and 0.9779. The performance of the adaptive DASO-based deep RNN was improved by 4.94251%, 6.2823%, 9.64145%, and 0.7154% when compared to state-of-the-art methods such as DBN, DSAE, CNN, and DASO-based deep RNN.

Figure 3c shows how the precision of training data was tested. With 80% of the training data, the created adaptive DASO-based deep RNN gets a specificity value of 0.9754. Existing methods such as DBN, DSAE, CNN, and DASO-based deep RNN get specificities of 0.7394, 0.8969, 0.9174, and 0.9611. The performance of the developed

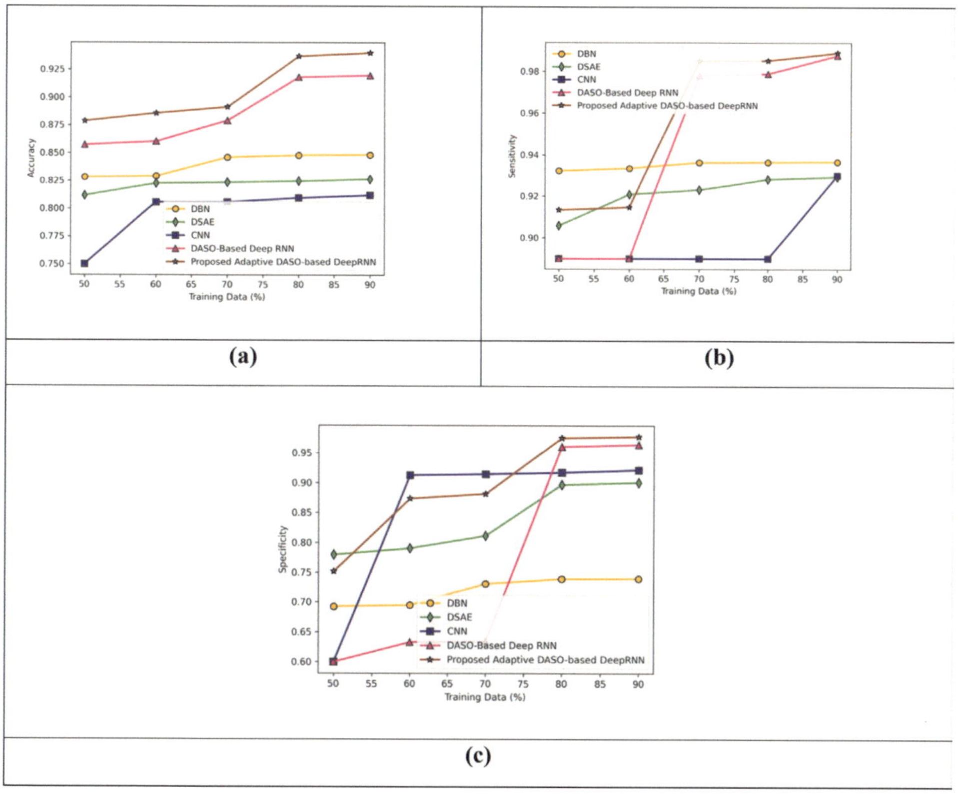

Figure 3.
Comparative analysis using dataset-1, (a) accuracy, (b) sensitivity, and (c) specificity.

adaptive DASO-based deep RNN was found to be better than state-of-the-art methods such as DBN, DSAE, CNN, and DASO-based deep RNN by 24,193%, 80,476%, 59,513%, and 14,712%, respectively.

5.4.2 Analysis using dataset-2

Figure 4a shows how the accuracy of the training data was compared to the accuracy of the test data. For 60% of the training data, the accuracy of the adaptive model is 0.9767, while the accuracy of DBN, DSAE, CNN, and DASO-based DRNN is 0.9305, 0.9329, 0.9388, and 0.956087, respectively. When comparing the developed adaptive model to state-of-the-art methods such as DBN, DSAE, CNN, and DASO-based deep RNN, the performance improvement is 4.7341%, 4.4860%, 3.8829%, and 2.1169%, respectively.

Figure 4b shows how the sensitivity analysis is done with the training data. For 70% of training data, the developed adaptive DASO-based deep RNN gets a specificity value of 0.9894, while existing methods such as DBN, DSAE, and CNN get values of 0.9560, 0.9238, 0.9280, and 0.9821 for sensitivity. When comparing the developed adaptive DASO-based deep RNN with the most advanced methods, such as DBN, DSAE, CNN, and DASO-based deep RNN, the performance improvement is 3.3776%, 6.6318%, 6.2063%, and 0.7416%, respectively.

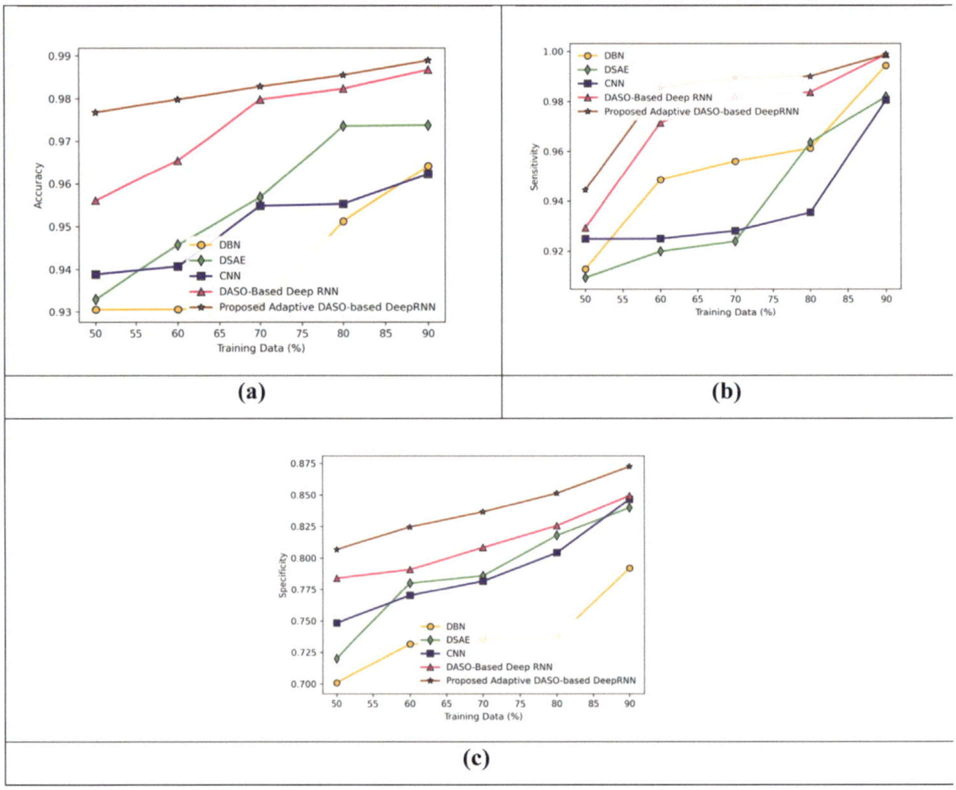

Figure 4.
Comparative analysis of dataset-2, (a) accuracy, (b) sensitivity, and (c) specificity.

Figure 4c shows how sensitivity is tested and how training data is used. When the training data is 80%, the developed adaptive model has a sensitivity of 0.8513. On the other hand, existing methods like DBN, DSAE, CNN, and DASO-based deep RNN have specificities of 0.7370, 0.8178, 0.8041, and 0.8255. When comparing the developed adaptive DASO-based deep RNN with the most advanced methods, like DBN, DSAE, CNN, and DASO-based deep RNN, the performance improvement was found to be 13.4295%, 3.9380%, 5.539%, and 3.02870%, respectively.

5.5 Comparative discussion

Table 1 shows a comparison of the adaptive model that has been created. Using dataset-1 as an example, the accuracy of the current DBN, DSAE, CNN, and DASO-based deep RNN is 0.8479, 0.8245, 0.8094, and 0.9180, while the accuracy of the proposed adaptive model is 0.93679, which is better. With dataset-1, the DBN, DSAE, CNN, and DASO-based deep RNN each got a sensitivity of 0.9364, 0.9281, 0.89, and 0.9788, but the suggested adaptive model did better, getting a sensitivity of 0.9851. With dataset-2, the accuracy of the existing DBN, DSAE, CNN, and DASO-based DRNN is 0.9512, 0.9735, 0.9552, and 0.9822, respectively, while the accuracy of the suggested adaptive model is 0.9854. With dataset-2, the specificity of the existing DBN, DSAE, CNN, and DASO-based deep RNN is 0.7370, 0.8178, 0.8041, and 0.82557, respectively, while the specificity of the suggested adaptive model is 0.8513.

Metrics/methods		DBN	DSAE	CNN	DASO-based deep RNN	Proposed adaptive model
dataset-1	Accuracy	0.847976	0.824557	0.809486	0.918039	0.936790
	Sensitivity	0.936468	0.928132	0.890000	0.978848	0.985166
	Specificity	0.739464	0.896963	0.917412	0.961113	0.975465
dataset-2	Accuracy	0.951206	0.973579	0.955279	0.982288	0.985469
	Sensitivity	0.961222	0.963585	0.935414	0.983639	0.990000
	Specificity	0.737022	0.817828	0.804192	0.825570	0.851355

Table 1.
Comparative discussion.

6. Conclusion

In this paper, a novel network ID mechanism named adaptive DASO-based deep RNN is developed to predict the abnormal behavior in the network. At first, the data are obtained from database and send this data to feature selection module using mutual information, which selects the relevant features. The features selected through feature selection are based on the threshold value. Once the features are selected, these features are forwarded to the IDS for predicting the malicious behavior in the network. The malicious activity is obtained by the developed DRNN, which is trained using Adaptive DASO algorithm. The Adaptive DASO model is designed by integrating adaptive concept, DE, and ASO. Although, the combined DA and ASO algorithm provides better result, but this method consumes high computational time. Thus, the adaptive concept is introduced with the DASO for reducing computational time. This algorithm predicts that the behavior of the network is either normal or abnormal. The weights are accurately measured by the developed Adaptive DASO algorithm through fitness function. In addition, the developed Adaptive DASO achieved optimal performance utilizing the evaluation metrics such as accuracy, sensitivity, and specificity with the values of 0.9854, 0.99, and 0.8513, using dataset-1. In the future, the detecting capacity of IDS can be enhanced by using some other optimization techniques.

Author details

Surendra Bhosale[1*], Achala Deshmukh[2], Bhushan Deore[1,3] and Parag Bhosale[4]

1 Veermata Jijabai Institute of Technology, Mumbai, India

2 Sinhgad College of Engineering, Pune, India

3 Ramrao Adik Institute of Technology, Navi Mumbai, India

4 George Mason University, Virginia, USA

*Address all correspondence to: sjbhosale@ee.vjti.ac.in

References

[1] Shone N, Ngoc TN, Phai VD, Shi Q. A deep learning approach to network intrusion detection. IEEE Transactions on Emerging Topics in Computational Intelligence. 2018;**2**(1):41-50

[2] Yin C, Zhu Y, Fei J, He X. A deep learning approach for intrusion detection using recurrent neural networks. IEEE Access. 2017;**5**:21954-21961

[3] Azad C, Jha VK. Fuzzy min–max neural network and particle swarm optimization based intrusion detection system. Microsystem Technologies. 2017;**23**(4):907-918

[4] Sohi SM, Seifert JP, Ganji F. RNNIDS: Enhancing network intrusion detection systems through deep learning. Computers and Security. 2020;**2020**:102151

[5] Mighan SN, Kahani M. A novel scalable intrusion detection system based on deep learning. International Journal of Information Sccurity. 2021;**20**(3):387-403

[6] Andresini G, Appice A, Malerba D. Autoencoder-based deep metric learning for network intrusion detection. Information Sciences. 2021;**569**:706-727

[7] Kaja N, Shaout A, Ma D. An intelligent intrusion detection system. Applied Intelligence. 2019;**49**(9):3235-3247

[8] Jin D, Lu Y, Qin J, Cheng Z, Mao Z. SwiftIDS: Real-time intrusion detection system based on LightGBM and parallel intrusion detection mechanism. Computers & Security. 2020;**97**:101984

[9] Sarker IH, Abushark YB, Alsolami F, Khan AI. Intrudtree: A machine learning based cyber security intrusion detection model. Symmetry. 2020;**12**(5):754

[10] Injadat M, Moubayed A, Nassif AB, Shami A. Multi-stage optimized machine learning framework for network intrusion detection. IEEE Transactions on Network and Service Management. 2020;**18**(2):1803-1816

[11] Bertoli GDC, Júnior LAP, Saotome O, Dos Santos AL, Verri FAN, Marcondes CAC, et al. An end-to-end framework for machine learning-based network intrusion detection system. IEEE Access. 2021;**9**:106790-106805

[12] Jiang K, Wang W, Wang A, Wu H. Network intrusion detection combined hybrid sampling with deep hierarchical network. IEEE Access. 2020;**8**:32464-32476

[13] Çavuşoğlu Ü. A new hybrid approach for intrusion detection using machine learning methods. Applied Intelligence. 2019;**49**(7):2735-2761

[14] Tang TA, Mhamdi L, McLernon D, Zaidi SAR, Ghogho M. Deep learning approach for network intrusion detection in software defined networking. In: Proceedings of 2016 International Conference on Wireless Networks and Mobile Communications (WINCOM). Fez, Morocco: IEEE; 2016. pp. 258-263

[15] Wu K, Chen Z, Li W. A novel intrusion detection model for a massive network using convolutional neural networks. IEEE Access. 2018;**6**:50850-50859

[16] Vinayakumar R, Alazab M, Soman KP, Poornachandran P, Al-Nemrat A, Venkatraman S. Deep learning approach for intelligent

intrusion detection system. IEEE Access. 2019;7:41525-41550

[17] Gao X, Shan C, Hu C, Niu Z, Liu Z. An adaptive ensemble machine learning model for intrusion detection. IEEE Access. 2019;7:82512-82521

[18] Otoum S, Kantarci B, Mouftah HT. On the feasibility of deep learning in sensor network intrusion detection. IEEE Networking Letters. 2019;**1**(2):68-71

[19] Yang H, Qin G, Ye L. Combined wireless network intrusion detection model based on deep learning. IEEE Access. 2019;7:82624-82632

[20] Wu P, Guo H. LuNET: A deep neural network for network intrusion detection. In: Proceedings of 2019 IEEE Symposium Series on Computational Intelligence (SSCI). Xiamen, China: IEEE; 2019. pp. 617-624

[21] Khan FA, Gumaei A, Derhab A, Hussain A. A novel two-stage deep learning model for efficient network intrusion detection. IEEE Access. 2019;7: 30373-30385

[22] Sohi SM, Seifert JP, Ganji F. RNNIDS: Enhancing network intrusion detection systems through deep learning. Computers & Security. 2021;**102**:102151

[23] Zeng Y, Gu H, Wei W, Guo Y. Deep-full-range: A deep learning based network encrypted traffic classification and intrusion detection framework. IEEE Access. 2019;7:45182-45190

[24] Jouad M, Diouani S, Houmani H, Zaki A. Security challenges in intrusion detection. In: Proceedings of International Conference on Cloud Technologies and Applications (CloudTech). Marrakech, Morocco. 2015. pp. 1-11

[25] Borkar GM, Mahajan AR. A secure and trust based on-demand multipath routing scheme for self-organized mobile ad-hoc networks. Wireless Networks. 2017;**23**(8):2455-2472

[26] Zhao W, Wang L, Zhang Z. A novel atom search optimization for dispersion coefficient estimation in groundwater. Future Generation Computer Systems. 2019;**91**:601-610

[27] Inoue M, Inoue S, Nishida T. Deep recurrent neural network for mobile human activity recognition with high throughput. Artificial Life and Robotics. 2018;**23**(2):173-185

[28] Erik G. Entropy and Mutual Information. Amherst; 2013

[29] Wu K, Chen Z, Li W. A novel intrusion detection model for a massive network using convolutional neural networks. IEEE Access. 2018;**2018**:1-1

[30] Khan FA, Gumaei A, Derhab A, Hussain A. TSDL: A two stage deep learning model for efficient network intrusion detection. IEEE Access. 2019:1-1

[31] Dong B, Wang X. Comparison deep learning method to traditional methods using for network intrusion detection. In: Proceedings of 8th IEEE International Conference on Communication Software and Networks (ICCSN). Beijing, China. 2016. pp. 581-585

[32] Sangeetha S, Ramya R, Dharani MK, Sathya P. Signature based semantic intrusion detection system on cloud. Information Systems Design and Intelligent Applications. 2015;**2015**:657-666

[33] NSL-KDD Dataset. Available from: https://www.unb.ca/cic/datasets/nsl.html [Accessed: August 2022]

[34] BoT-IoT Dataset. Available from: https://research.unsw.edu.au/projects/BoT-IoT-dataset [Accessed: August 2022]

Chapter 6

Anomaly Detection in Intrusion Detection Systems

Siamak Parhizkari

Abstract

Intrusion detection systems (IDS) play a critical role in network security by monitoring systems and network traffic to detect anomalies and attacks. This study explores the different types of IDS, including host-based and network-based, along with their deployment scenarios. A key focus is on incorporating anomaly detection techniques within IDS to identify novel and unknown threats that evade signature-based methods. Statistical approaches like outlier detection and machine learning techniques like neural networks are discussed for building effective anomaly detection models. Data collection and preprocessing techniques, including feature engineering, are examined. Both unsupervised techniques like clustering and density estimation and supervised methods like classification are covered. Evaluation datasets and performance metrics for assessing anomaly detection models are highlighted. Challenges like curse of dimensionality and concept drift are outlined. Emerging trends include integrating deep learning and explainable AI into anomaly detection. Overall, this comprehensive study examines the role of anomaly detection within IDS, delves into various techniques and algorithms, surveys evaluation practices, discusses limitations and challenges, and provides insights into future research directions to advance network security through improved anomaly detection capabilities.

Keywords: anomaly detection, intrusion detection systems (IDS), fraud detection, cybersecurity, abnormal patterns

1. Introduction

An intrusion detection system (IDS) is a security tool designed to monitor network or system activities to detect and respond to unauthorized or malicious activities. It serves as an additional layer of defense in a comprehensive cybersecurity strategy.

The primary goal of an IDS is to identify and alert security administrators about potential security incidents, such as unauthorized access attempts, malware infections, or suspicious network traffic patterns. By analyzing network packets, log files, system activities, and other relevant data, IDS can help detect and respond to security threats in real-time.

There are two main types of IDS:

Network-based intrusion detection systems (NIDS) [1–7]: NIDS monitors network traffic in real-time, analyzing packets to identify suspicious or malicious activity. It operates at the network layer and can detect threats such as port scanning, denial-of-service (DoS) attacks, and network intrusions. NIDS can be deployed as a standalone device or as part of a network security infrastructure.

Host-based intrusion detection systems (HIDS) [1–7]: HIDS monitors the activities occurring on individual hosts or endpoints, such as servers or workstations. It analyzes system logs, file integrity, and user activities to identify unauthorized access attempts, privilege escalations, or suspicious behavior at the host level. HIDS is particularly useful for detecting insider threats or malware infections that may bypass network-based defenses.

IDS employs different detection techniques to identify potential threats:

Signature-based detection [4]: This technique relies on a database of known attack signatures or patterns. IDS compares the incoming network traffic or system activities against these signatures to identify known attacks. While effective against known threats, signature-based detection may struggle with detecting new or zero-day attacks.

Anomaly-based detection [4–6, 8]: Anomaly detection involves establishing a baseline of normal behavior for a network or system and then identifying deviations from this baseline. It analyzes traffic patterns, system performance, user behavior, and other metrics to detect anomalies that could indicate a potential security breach.

When an IDS detects an intrusion or suspicious activity, it generates an alert or notification for security administrators. These alerts provide information about the nature of the incident, the affected system or network, and any additional details to aid in the response and mitigation process.

It is important to note that IDS is not a standalone solution but works in conjunction with other security measures like firewalls, antivirus software, and security policies. Additionally, intrusion prevention systems (IPS) are often used in conjunction with IDS to not only detect but also actively block or prevent detected threats.

In summary, Intrusion Detection Systems play a crucial role in identifying and responding to potential security incidents in real-time. By monitoring network and system activities, IDS helps organizations strengthen their overall security posture and minimize the potential impact of cyber threats.

2. Anomaly detection techniques in IDS

Table 1 shows Anomaly detection techniques with pros and cons.

2.1 Signature-based detection vs. anomaly detection

Signature-based detection, also known as rule-based detection, relies on predefined signatures or patterns of known attacks to identify intrusions. However, signature-based detection has limitations as it can only detect known attacks for which signatures have been defined. New or unknown attacks can easily evade signature-based detection. Anomaly detection techniques, on the other hand, focus on identifying deviations from normal behavior, without relying on predefined signatures. This makes anomaly detection more effective in detecting unknown or novel attacks that do not have specific signatures [4, 6, 9]. **Figure 1** shows the concept of signature-based IDS.

Anomaly Detection Techniques in IDS			
Techniques	**Models**	**Pros**	**Cons**
Signature-based	• Pattern matching • Protocol analysis • Content inspection • Log analysis	• High accuracy for known attacks • Low false alarm rate • Easy deployment • Low computational overhead	• Inability to detect New or unknown attacks • Dependency on signature update • Lack of flexibility • Limited coverage
Statistical-based	• Outlier detection • Time series analysis • Statistical Modeling	• Well-established method with a solid theoretical foundation • Suitable for detection simple anomalies • Interpretable results	• Limited ability to detect complex or sophisticated anomalies • Sensitivity to data distribution and assumptions • Difficulties in handling high-dimensional data
Machine Learning	• Clustering • Classification • Neural Networks	• Ability to handle complex and non-linear patterns • Effective for identifying subtle anomalies • Adaptability to changing environments • Can learn from unlabeled or partially labeled data	• Requirement of large labeled training datasets • Overfitting if no properly tuned or validated • Computationally intensive for complex algorithms • Black-box nature may lack interpretability
Hybrid Approaches	Statistical + Machine learning methods	• Leveraging the strengths of both statistical and machine learning techniques • Improved detection accuracy and robustness • Enhanced ability to handle diverse anomalies	• Increase complexity and potential for integration challenges • Higher computational requirements • Potential trade-off between interpretability and performance

Table 1.
Anomaly detection techniques with pros and cons.

2.2 Statistical approaches for anomaly detection

Statistical approaches are commonly employed for anomaly detection in IDS. These techniques involve the use of statistical methods to establish normal behavior baselines and detect deviations from these baselines. Outlier detection algorithms, such as the statistical outlier detection method or the Z-score method, are used to identify data points that significantly deviate from the expected behavior. Time series analysis techniques, such as autoregressive integrated moving average (ARIMA) models [10, 11], are used to detect anomalies in temporal data. Statistical modeling approaches, such as Gaussian mixture models or hidden Markov models, are utilized to capture the statistical characteristics of normal behavior and detect anomalies based on deviations from the learned models [4, 5].

2.3 Machine learning approaches for anomaly detection

Machine learning algorithms play a crucial role in anomaly detection for IDS. These algorithms can learn patterns and behaviors from historical data and apply that knowledge to detect anomalies in real-time. Clustering algorithms, such as k-means or

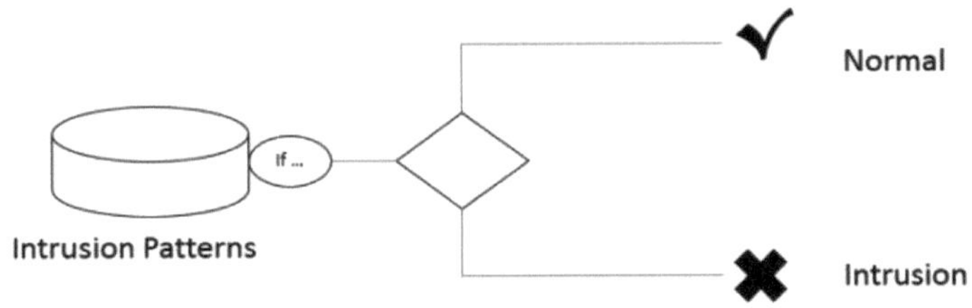

Figure 1.
Concept of signature based IDS [4].

DBSCAN, group similar instances together and flag instances that do not fit into any cluster as anomalies. Classification algorithms, such as support vector machines (SVM) or random forests, learn from labeled data to classify instances as normal or anomalous. Neural networks, including deep learning models like convolutional neural networks (CNN) [9] or recurrent neural networks (RNN) [9], can capture complex patterns and relationships to identify anomalies [12]. **Figure 2** shows machine learning approaches in IDS.

2.4 Hybrid approaches

Hybrid approaches combine both statistical and machine learning techniques to improve the accuracy and effectiveness of anomaly detection in IDS [4]. By leveraging the strengths of different approaches, hybrid models can provide enhanced detection capabilities. For example, a hybrid approach may use statistical techniques to establish baseline behavior and machine learning algorithms to classify instances as normal or anomalous. This combination allows for a more comprehensive and robust anomaly detection system.

3. Data collection and preprocessing in IDS

Data sources for IDS [4, 12].

Intrusion detection systems (IDS) rely on various sources of data to detect anomalies and potential security breaches. Some common data sources used in IDS include:

1. Network traffic logs: IDS can analyze network traffic logs to monitor incoming and outgoing network packets, protocols used, source and destination IP addresses, ports, and other relevant information. Network traffic logs provide valuable insights into communication patterns and can help to detect anomalies such as unusual traffic volumes, suspicious connections, or protocol violations.

2. System logs: System logs record events and activities within the operating system or specific applications. IDS can analyze system logs to identify abnormal system behavior, such as unauthorized access attempts, changes to system configurations, or unexpected system errors. System logs may include information about login attempts, file access, process execution, or resource utilization.

3. Audit trails: Audit trails capture detailed information about user activities and actions within a system. They record events such as file access, privilege changes,

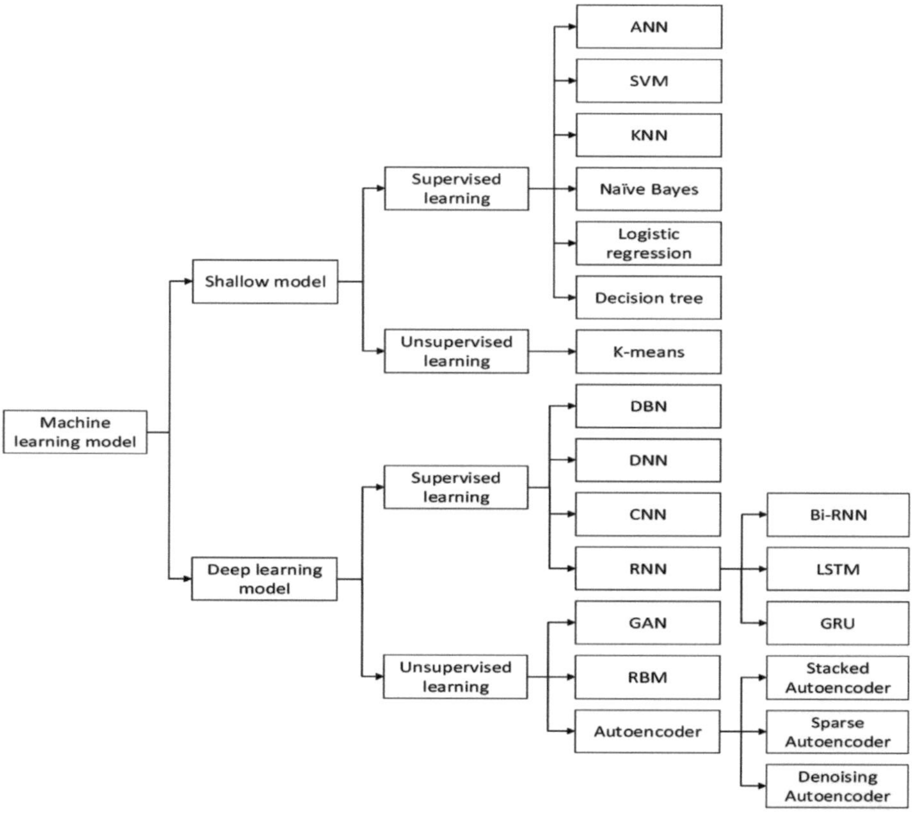

Figure 2.
Machine learning approaches in IDS [12].

user authentication, or administrative actions. Analyzing audit trails can help to identify unauthorized actions, unusual user behavior, or privilege escalation attempts.

Data preprocessing techniques.

Data preprocessing [13, 14] plays a crucial role in preparing the data for effective anomaly detection in IDS. Several techniques are commonly used in the preprocessing stage, including:

1. Data cleaning [14]: Data cleaning involves removing or correcting inconsistent, irrelevant, or noisy data. This process may include handling missing values, dealing with outliers, and resolving inconsistencies in the data. Cleaning the data helps to ensure the quality and reliability of the input data for anomaly detection.

2. Feature selection [4, 15]: Feature selection aims to identify the most relevant and informative features for anomaly detection. In IDS, this involves selecting the attributes or variables that provide the most discriminative information about

normal and anomalous behavior. Feature selection can help to reduce computational complexity, improve detection accuracy, and eliminate redundant or irrelevant features.

3. Normalization [14]: Normalization is the process of scaling data to a common range or distribution. It ensures that different features are on a comparable scale, which is essential for certain anomaly detection algorithms that rely on distance or similarity measures. Normalization techniques include min-max scaling, z-score normalization, or logarithmic transformations.

4. Dimensionality reduction [16]: Dimensionality reduction techniques aim to reduce the number of features while preserving the most important information. High-dimensional data can be computationally expensive and prone to overfitting. Techniques such as principal component analysis (PCA), linear discriminant analysis (LDA), or t-distributed stochastic neighbor embedding (t-SNE) can help to reduce the dimensionality of the data while retaining its essential characteristics.

4. Unsupervised anomaly detection in IDS

Unsupervised anomaly detection techniques in intrusion detection systems (IDS) aim to identify anomalies in data without relying on pre-labeled instances of normal and anomalous behavior [4, 9]. These techniques are particularly useful in scenarios where labeled training data is scarce or unavailable, making it challenging to train supervised models. Unsupervised anomaly detection methods utilize statistical, clustering, or density-based approaches to identify patterns that deviate from normal behavior. Here are some commonly used unsupervised anomaly detection techniques in IDS and **Figure 3** shows a summary of these techniques:

1. Statistical-based techniques: Statistical-based techniques are commonly used for unsupervised anomaly detection in intrusion detection systems (IDS). As we said before, these techniques analyze the statistical properties of the data to identify instances that deviate significantly from the expected behavior. The underlying assumption is that normal behavior follows a certain statistical distribution, and any deviation from this distribution is considered anomalous. Here are some commonly used statistical-based techniques:

    - Gaussian distribution: The Gaussian distribution, also known as the normal distribution, is frequently used in statistical-based anomaly detection. It assumes that the normal behavior of the data follows a bell-shaped curve. Anomalies are identified as instances that fall outside a specified range or threshold based on the estimated mean and standard deviation of the data. Instances that lie in the tails of the distribution, beyond a certain number of standard deviations from the mean, are considered anomalies.

    - Mahalanobis distance: The Mahalanobis distance measures the distance between a data point and the center of a distribution, taking into account the

DOI: http://dx.doi.org/10.5772/intechopen.112733

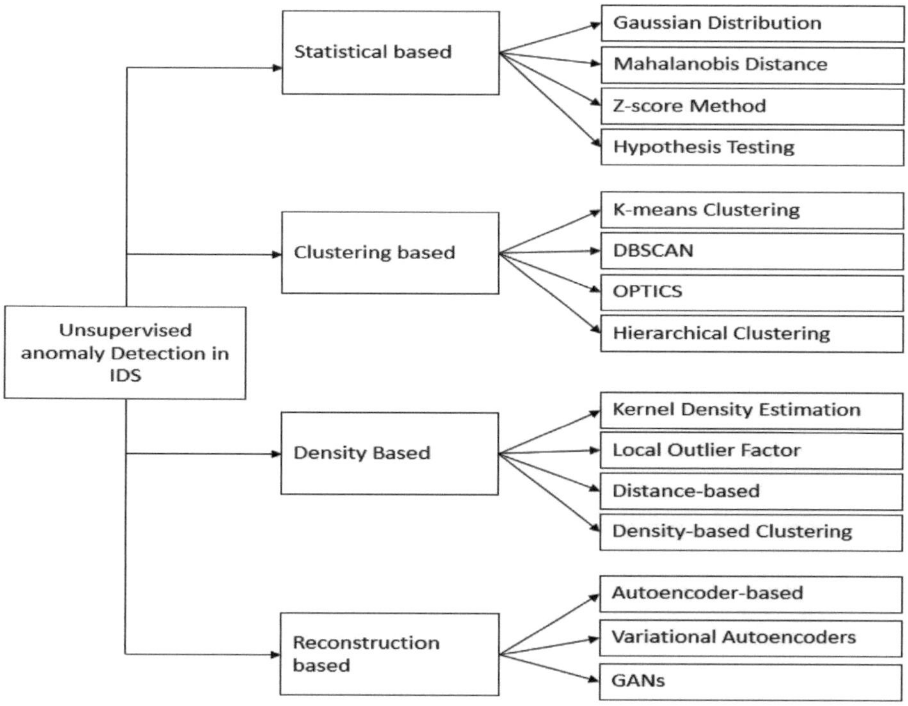

Figure 3.
Summarize of unsupervised techniques.

correlation between variables. It accounts for the covariance structure of the data and is particularly useful when the variables are correlated. The Mahalanobis distance can be used to detect anomalies by comparing the distance of each data point to a threshold value. Points with a large Mahalanobis distance are considered anomalies.

- Z-score method: The Z-score method is a simple statistical technique for anomaly detection. It calculates the standard deviation from the mean for each data point and expresses it as a Z-score. The Z-score represents the number of standard deviations a data point is away from the mean. Anomalies are identified as data points with a Z-score exceeding a specified threshold. This method is particularly useful when the data is normally distributed.

- Hypothesis testing: Hypothesis testing is a statistical technique used to determine the likelihood that an observed deviation from the expected behavior is due to chance or represents an anomaly. Commonly used hypothesis tests include the t-test, chi-square test, or Kolmogorov-Smirnov test. These tests compare the observed data to a reference distribution or expected behavior and calculate a p-value. If the p-value is below a predefined significance level, the deviation is considered significant, and the instance is flagged as an anomaly.

Statistical-based techniques provide a solid foundation for detecting anomalies based on deviations from expected statistical behavior. However, it is important to note that these methods assume the data follows specific statistical distributions and may not be suitable for data with complex or non-parametric distributions. Additionally, choosing appropriate thresholds or significance levels is crucial and requires careful consideration and domain knowledge.

2. Clustering-based techniques: Clustering-based techniques as shown in **Figure 4** are commonly used for unsupervised anomaly detection in intrusion detection systems (IDS). These techniques aim to partition the data into clusters based on the similarity or density of instances [14, 17]. Anomalies are identified as instances that do not belong to any cluster or are located far from the clusters. Here are some commonly used clustering-based techniques:

   - K-means clustering: K-means clustering is a popular technique that aims to partition the data into K clusters. The algorithm iteratively assigns data points to the nearest cluster centroid based on distance measures such as Euclidean distance. Anomalies are typically identified as instances that do not fit well into any cluster or are located far from the cluster centroids. However, K-means alone may not be sufficient for anomaly detection as it assumes that all clusters have similar sizes and shapes, which may not hold true for anomalous instances.

   - Density-based spatial clustering of applications with noise (DBSCAN): DBSCAN is a density-based clustering algorithm that identifies clusters based on the density of instances [18]. It groups together instances that are close to each other and have a sufficient number of nearby neighbors. Anomalies are typically instances that do not have enough nearby neighbors to form a cluster and are considered noise points. DBSCAN can effectively identify clusters of different shapes and sizes, making it suitable for detecting anomalies that do not conform to regular cluster patterns.

   - Ordering points to identify the clustering structure (OPTICS): OPTICS is an extension of DBSCAN that provides a hierarchical view of the clustering structure. It orders instances based on their density and identifies core

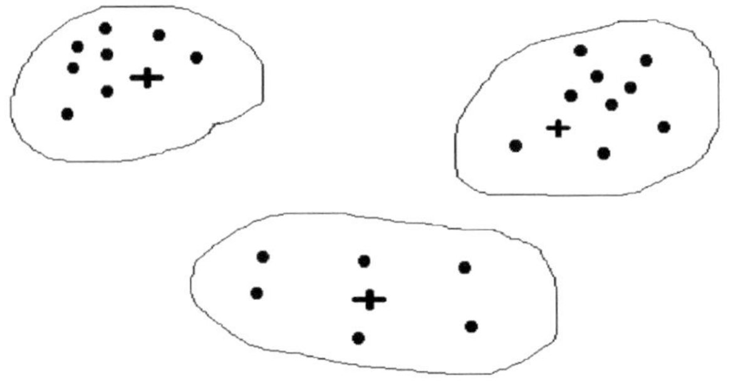

Figure 4.
Clustering [14].

points, reachability distances, and clusters. Anomalies are typically instances that have low density and are located in regions with sparse or no clusters. OPTICS allows for flexible parameterization, making it more adaptive to different datasets and providing a richer characterization of the data structure.

- Hierarchical clustering: Hierarchical clustering methods create a hierarchy of clusters by successively merging or splitting clusters based on their similarity. Agglomerative hierarchical clustering starts with each instance as a separate cluster and iteratively merges similar clusters until a single cluster is formed. Divisive hierarchical clustering starts with all instances in one cluster and iteratively splits the cluster into smaller clusters. Anomalies can be identified as instances that do not fit well into any cluster or do not conform to the hierarchical structure.

Clustering-based techniques offer flexibility in detecting anomalies by identifying instances that do not conform to regular cluster patterns. However, these techniques require careful consideration of parameters such as the number of clusters or density thresholds, and the interpretation of anomalies may depend on the dataset and the clustering algorithm used.

3. Density-based techniques: Density-based techniques are commonly used for unsupervised anomaly detection in intrusion detection systems (IDS). These techniques focus on estimating the density distribution of the data and identify anomalies as instances that lie in regions of low density [19]. Here are some commonly used density-based techniques:

- Kernel density estimation (KDE): Kernel density estimation is a non-parametric technique used to estimate the underlying density distribution of the data. It places a kernel function on each data point and sums them to estimate the density at any given point. Anomalies are typically identified as instances with significantly lower density values compared to the majority of the data. The choice of kernel function and bandwidth parameter affects the smoothness and accuracy of density estimation.

- Local outlier factor (LOF): The local outlier factor measures the deviation of an instance's density compared to its neighboring instances. It calculates a local density for each data point based on the distances to its k nearest neighbors. Anomalies are identified as instances with significantly lower local densities compared to their neighbors. LOF takes into account the local density variations in the data, making it robust to varying densities and useful for detecting anomalies in clusters or regions of different densities.

- Distance-based techniques: Distance-based density estimation techniques measure the distances between instances and identify anomalies based on deviations from the expected distance distribution. For example, the nearest neighbor distance (NND) approach calculates the average distance to the k nearest neighbors for each instance. Anomalies are identified as instances with significantly larger or smaller distances compared to the majority of the data. Distance-based techniques are effective in identifying anomalies that exhibit unusual distance patterns.

- Density-based clustering [18]: Density-based clustering algorithms, such as DBSCAN, can also be used for anomaly detection. These algorithms identify clusters based on the density of instances and label as anomalies the instances that do not belong to any cluster. Anomalies are typically located in regions with low density or as individual points far from the clusters.

Density-based techniques provide flexibility in detecting anomalies by focusing on regions of low density or deviations from expected distance patterns. These techniques are effective in identifying anomalies that do not conform to regular density distributions or exhibit unusual distance patterns. However, careful parameter selection, such as the neighborhood size or density thresholds, is important to ensure accurate anomaly detection.

4. Reconstruction-based techniques [20]: Reconstruction-based techniques are a class of anomaly detection techniques that aim to reconstruct the normal behavior of the data and identify anomalies based on the errors or deviations from this reconstruction. These techniques typically employ autoencoders shown in **Figure 5** or similar models to learn the underlying patterns in the data and use them to reconstruct or generate the data. Here are some commonly used reconstruction-based techniques:

   - Autoencoder-based anomaly detection: Autoencoders are neural network models that are trained to reconstruct their input data. They consist of an encoder that compresses the input data into a lower-dimensional representation and a decoder that reconstructs the data from this representation. During training, autoencoders learn to minimize the reconstruction error by capturing the patterns and regularities in the data. Anomalies are identified as instances that result in high reconstruction errors, indicating deviations from the learned normal behavior.

   - Variational autoencoders (VAEs): Variational autoencoders are a type of generative model that learns a low-dimensional representation of the data and generates new samples by sampling from this learned representation. VAEs consist of an encoder that learns the parameters of a probability distribution in the latent space and a decoder that generates samples from this distribution. Anomalies can be identified based on the reconstruction error or by measuring the dissimilarity between the original data and the

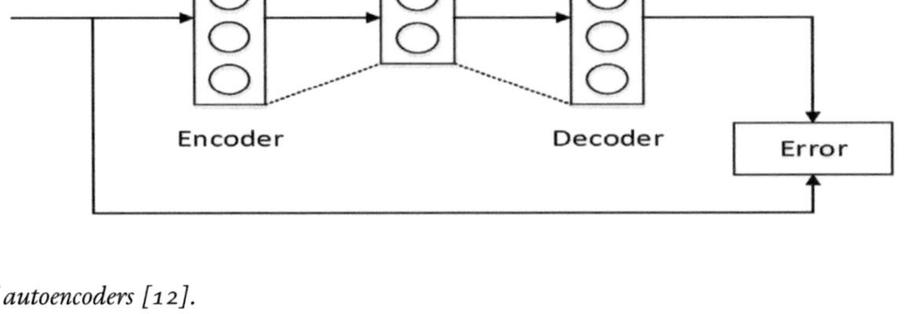

Figure 5.
Structure of autoencoders [12].

generated samples. VAEs can capture the underlying distribution of the data and detect anomalies that deviate significantly from this distribution.

- Generative adversarial networks (GANs) [21]: Generative adversarial networks are another type of generative model that consists of a generator network and a discriminator network. The generator network learns to generate realistic samples that resemble the normal behavior of the data, while the discriminator network learns to distinguish between real and generated samples. Anomalies can be identified as instances that are not well captured by the generator network or are classified as fake by the discriminator network. GANs can learn complex data distributions and detect anomalies that differ significantly from the learned distribution.

Reconstruction-based techniques offer the advantage of learning the normal behavior of the data and identifying anomalies based on deviations from this learned behavior. They can capture complex patterns and variations in the data, making them effective for detecting anomalies that do not conform to specific statistical or density distributions. However, these techniques require a representative dataset of normal behavior for training the models and may be sensitive to the choice of model architecture and training parameters.

5. Supervised anomaly detection in IDS

Supervised anomaly detection in intrusion detection systems (IDS) involves training a model on labeled data, where both normal and anomalous instances are explicitly identified [4, 9]. The model learns the patterns and characteristics of normal behavior during the training phase and can subsequently classify new instances as either normal or anomalous based on the learned knowledge. Here are some commonly used techniques for supervised anomaly detection in IDS:

1. Supervised machine learning algorithms [4]: Supervised machine learning algorithms, such as decision trees [4], random forests, support vector machines (SVM), and neural networks, can be applied to train models for supervised anomaly detection in IDS. These algorithms learn from labeled data, where the labels indicate whether an instance is normal or anomalous. The trained models can then classify new instances as either normal or anomalous based on the learned patterns and decision boundaries.

2. Ensemble methods [4]: Ensemble methods combine multiple models to improve the accuracy and robustness of supervised anomaly detection. Techniques such as bagging, boosting, and stacking can be employed to create an ensemble of models that collectively make predictions. Each individual model may use a different algorithm or have different parameter settings, and the final prediction is often determined through voting or averaging. Ensemble methods can effectively handle complex data distributions and improve the overall detection performance.

3. Deep learning [9]: Deep learning techniques, such as convolutional neural networks (CNNs) shown in **Figure 6**, recurrent neural networks (RNNs) shown

in **Figure 7**, and deep autoencoders, have shown promising results in supervised anomaly detection. These models can learn complex representations of the input data, capture intricate patterns, and generalize well to unseen instances. Deep learning approaches require large amounts of labeled data and can be computationally intensive but can achieve high accuracy in detecting anomalies in IDS.

4. Feature engineering: Feature engineering plays a crucial role in supervised anomaly detection. It involves selecting relevant features from the data or designing new features that can effectively discriminate between normal and anomalous instances. Domain knowledge and expertise are often employed to identify informative features that can capture the distinguishing characteristics of anomalies. Feature engineering techniques, such as dimensionality reduction, feature selection, and feature transformation, can improve the detection performance of supervised anomaly detection models.

5. One-class support vector machines (SVM): One-class support vector machines (SVM) is a popular technique for anomaly detection that falls under the category of supervised learning. Unlike traditional SVMs that are designed for binary classification tasks, One-class SVM is specifically designed for the task of learning a model of normal data and identifying anomalies based on deviations from this model [22]. Here's how One-class SVM works:

    - Training phase: In the training phase, One-class SVM learns a decision boundary that encloses the majority of the training data points, representing the normal class. The goal is to find a hyperplane that maximally separates the normal data instances from the origin or the center of the feature space.

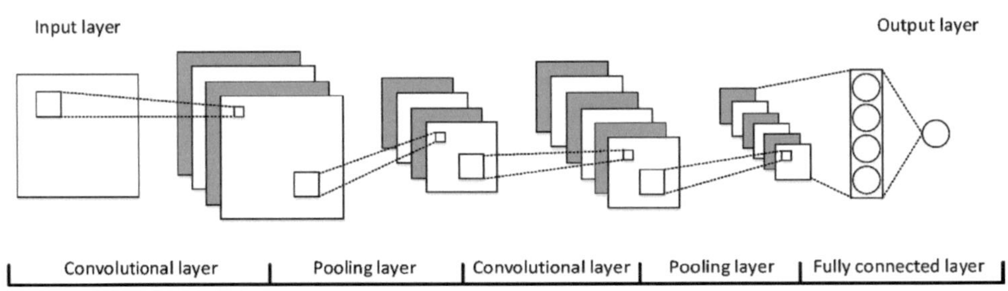

Figure 6.
Structure of convolutional neural network [12].

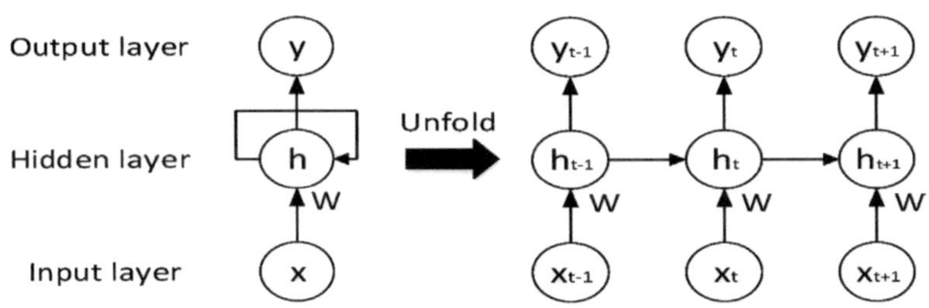

Figure 7.
Structure of recurrent neural network [12].

- Decision function: Once the model is trained, the decision function of the One-class SVM is used to classify new instances as either normal or anomalous. The decision function assigns a score or distance to each instance, indicating how well it fits within the learned model. Instances with positive scores are considered normal, while instances with negative scores are classified as anomalies.

- Kernel trick: One-class SVM can make use of the kernel trick to handle nonlinear data distributions. By mapping the input data into a high-dimensional feature space, the One-class SVM can find a nonlinear decision boundary that better separates normal instances from anomalies. Commonly used kernel functions include the radial basis function (RBF) kernel and the polynomial kernel.

One-class SVM offers several advantages for anomaly detection:

- It can handle data with high-dimensional feature spaces effectively.

- It is robust to outliers in the training data.

- It can capture complex decision boundaries, including nonlinear ones, through the use of the kernel trick.

- It does not require labeled anomalies for training, as it focuses solely on learning the normal class.

However, One-class SVM also has certain limitations:

- It may struggle when the normal class exhibits significant variations or when the normal data distribution is not well-represented in the training set.

- It may be sensitive to the choice of kernel function and its hyperparameters, requiring careful tuning for optimal performance.

- It does not provide direct probabilistic outputs, making it challenging to interpret the anomaly scores as probability estimates.

Supervised anomaly detection in IDS offers the advantage of explicitly labeled data for training and can achieve high detection accuracy. However, it relies on the availability of accurately labeled training data and may face challenges when dealing with evolving or previously unseen anomalies. Moreover, supervised approaches may not capture novel or unknown anomalies that were not present in the training data.

6. Evaluation and performance metrics in IDS

Evaluation datasets play a crucial role in assessing the performance of anomaly detection techniques in intrusion detection systems (IDS). These datasets are used to evaluate how well a detection technique can accurately classify instances as normal or anomalous. Several datasets have been widely used in the field of IDS for evaluation purposes. Here are some commonly used datasets:

- KDD Cup 99 [4, 9]: The KDD Cup 99 dataset is one of the most widely used datasets for evaluating IDS techniques. It was created for the Third International Knowledge Discovery and Data Mining Tools Competition held in 1999. The dataset contains a large number of network connection records generated in a simulated network environment, with various types of attacks and normal traffic.
- NSL-KDD [4, 9]: The NSL-KDD dataset is an updated version of the original KDD Cup 99 dataset. It addresses some of the limitations and drawbacks of the original dataset, such as redundant and irrelevant features. NSL-KDD provides a more balanced and realistic representation of network traffic data, making it suitable for evaluating IDS techniques.
- CICIDS2017 [4, 23, 24]: The CICIDS2017 dataset is a recent dataset that was developed for evaluating IDS techniques in the context of real-world network traffic. It consists of network traffic data collected from a real network environment, containing various types of attacks and normal traffic instances. The dataset offers a comprehensive and diverse set of scenarios for evaluating IDS performance.

Performance metrics are used to quantitatively measure the effectiveness of anomaly detection techniques in IDS. These metrics provide insights into the model's accuracy, precision, recall, and overall performance. Here are some commonly used performance metrics:

1. Accuracy [4, 9, 25, 26]: Accuracy measures the overall correctness of the model's predictions. It calculates the ratio of correctly classified instances to the total number of instances. However, accuracy can be misleading in imbalanced datasets where anomalies are rare.

$$Accuracy = \frac{TP + TN}{TP + TN + FP + FN} \tag{1}$$

2. Precision [4, 9, 25, 26]: Precision measures the proportion of correctly identified anomalies among all instances classified as anomalies. It focuses on the correctness of positive predictions, indicating the model's ability to avoid false positives.

$$Precision = \frac{TP}{TP + FP} \tag{2}$$

3. Recall [4, 9, 25, 26]: Recall, also known as sensitivity or true positive rate, measures the proportion of actual anomalies that are correctly identified by the model. It represents the model's ability to detect anomalies and avoid false negatives.

$$Recall = \frac{TP}{TP + FN} \tag{3}$$

4. F1-score [4, 9, 25, 26]: The F1-score is a harmonic mean of precision and recall, providing a balanced measure of the model's performance. It considers both false positives and false negatives and is especially useful when dealing with imbalanced datasets.

$$f - score = \frac{Precision.Recall}{Precision + Recall} \tag{4}$$

5. ROC curve and AUC [4, 9]: The receiver operating characteristic (ROC) curve illustrates the trade-off between the true positive rate (TPR) and the false positive rate (FPR) at different classification thresholds. The area under the curve (AUC) summarizes the performance of the model across all possible thresholds. A higher AUC indicates better discrimination between normal and anomalous instances.

These performance metrics help in evaluating the accuracy, effectiveness, and reliability of anomaly detection techniques in IDS.

7. Challenges in anomaly detection for IDS

1. Curse of dimensionality [27]: The curse of dimensionality refers to the phenomenon where the effectiveness of certain algorithms and techniques deteriorates as the dimensionality of the data increases. In the context of intrusion detection systems (IDS), the curse of dimensionality poses a significant challenge for anomaly detection.

   Anomaly detection in IDS often involves analyzing high-dimensional data, such as network traffic logs, system logs, or audit trails. Each data instance is typically represented by a large number of features or attributes that describe various aspects of the network or system behavior. However, as the number of features increases, the available data becomes increasingly sparse in the high-dimensional space.

   The curse of dimensionality has several implications for anomaly detection in IDS:

   - Insufficient data: As the number of dimensions (features) increases, the amount of available data decreases exponentially. In other words, the data becomes sparse, and the number of instances representing normal and anomalous behavior becomes limited. This scarcity of data can lead to poor generalization and inaccurate anomaly detection.

   - Increased complexity: With a high number of dimensions, the complexity of the anomaly detection problem also increases. Anomaly detection algorithms may struggle to effectively capture patterns and relationships among the features, leading to decreased detection accuracy.

   - Increased computational cost [28]: The computational cost of processing high-dimensional data is significantly higher than processing data with a lower dimensionality. Anomaly detection algorithms may require more computational resources and time to analyze and classify instances accurately, affecting the real-time performance of IDS.

To mitigate the curse of dimensionality in IDS, various techniques can be employed:

- Dimensionality reduction: Dimensionality reduction methods aim to reduce the number of features while preserving the most relevant information. Techniques such as principal component analysis (PCA) and t-distributed stochastic neighbor embedding (t-SNE) can be used to reduce the dimensionality of the data, making it more manageable for anomaly detection algorithms.

- Feature selection: Feature selection techniques help to identify the most informative and discriminative features that contribute to anomaly detection. By selecting a subset of relevant features, the curse of dimensionality can be alleviated, reducing computational complexity and improving detection accuracy.

- Feature engineering: In addition to dimensionality reduction and feature selection, feature engineering involves transforming or creating new features that better represent the underlying characteristics of normal and anomalous behavior. This process can help to extract more meaningful information from the high-dimensional data and enhance the performance of anomaly detection algorithms.

Overall, addressing the curse of dimensionality in IDS requires careful consideration of data representation, feature selection, and dimensionality reduction techniques. By reducing the dimensionality of the data and focusing on relevant features, anomaly detection algorithms can be more effective in accurately identifying anomalies in high-dimensional data.

2. Concept drift [29, 30]: Concept drift refers to the phenomenon where the underlying data distribution, which defines what is considered normal or anomalous, changes over time. In the context of intrusion detection systems (IDS), concept drift poses a significant challenge for anomaly detection.

   In IDS, anomaly detection models are trained on historical data to learn patterns of normal behavior and identify deviations from those patterns as anomalies. However, the characteristics of network traffic and system behavior can evolve over time due to various factors such as changes in network infrastructure, software updates, and emerging attack techniques. As a result, the learned model may become outdated and less effective in detecting new types of anomalies.

   Concept drift in IDS can occur in different forms:

   - Gradual concept drift: In gradual concept drift, the change in the underlying data distribution is relatively slow and progressive. The statistical properties of the data gradually shift over time, leading to a gradual degradation in the

DOI: http://dx.doi.org/10.5772/intechopen.112733

performance of the anomaly detection model. This type of concept drift requires continuous monitoring and adaptation of the model to maintain its effectiveness.

- Sudden concept drift: In sudden concept drift, the change in the underlying data distribution occurs abruptly and unpredictably. This can happen due to sudden changes in network conditions, system configurations, or the introduction of new attack techniques. Sudden concept drift poses a significant challenge as the model needs to quickly adapt to the new data distribution to accurately detect anomalies.

Addressing concept drift in IDS is essential to maintain the effectiveness of anomaly detection over time. Several techniques can be employed:

- Online learning: Online learning approaches allow the anomaly detection model to continuously adapt to new data as it arrives. By updating the model with each new data point or in small batches, online learning can capture and respond to concept drift in real-time. Techniques such as incremental learning, ensemble methods, and adaptive models can be used to achieve online learning.

- Change detection: Change detection techniques monitor the statistical properties of the data and detect significant changes that indicate concept drift. By periodically comparing the current data distribution with the historical distribution, these methods can trigger model retraining or adaptation when a significant change is detected. Statistical methods like control charts, cumulative sum (CUSUM), and change point detection algorithms can be used for change detection.

- Ensemble methods: Ensemble methods combine multiple anomaly detection models or algorithms to improve detection performance and resilience to concept drift. By aggregating the decisions of multiple models, ensemble methods can adapt to changing data distributions and make more robust anomaly predictions. Techniques like ensemble averaging, boosting, and stacking can be applied to create ensemble models.

It is important to note that concept drift detection and adaptation in IDS is an ongoing research area, and the development of effective techniques to handle concept drift remains an active research topic.

3. Adversarial attacks [31–33]: Adversarial attacks in IDS refer to deliberate attempts by adversaries to exploit vulnerabilities in the system and manipulate its behavior in order to evade detection or cause misclassification of normal or malicious activities. These attacks are specifically designed to target the anomaly detection capabilities of IDS and can have serious consequences for the security of the network.

There are different types of adversarial attacks that can be launched against IDS:

- Evasion attacks: Evasion attacks aim to manipulate the input data in a way that the IDS fails to detect or correctly classify the malicious activities. Attackers carefully craft the input features to deceive the IDS into treating them as normal behavior, thus evading detection. Evasion attacks often involve carefully modifying or adding features to manipulate the decision boundary of the IDS.

- Poisoning attacks: Poisoning attacks occur during the training phase of the IDS and involve injecting malicious or manipulated data into the training set. By poisoning the training data, attackers aim to manipulate the learning process of the IDS, compromising its detection capabilities. The poisoned data can introduce biases or alter the statistical properties of the training set, leading to degraded performance or increased false positives/negatives.

- Stealth attacks: Stealth attacks aim to exploit the specific weaknesses or blind spots of the IDS to remain undetected. These attacks often involve carefully crafted sequences of activities that exploit temporal or contextual vulnerabilities, making it difficult for the IDS to identify them as anomalies. Stealth attacks can leverage timing patterns, bursty activities, or sophisticated evasion techniques to bypass detection.

- Data injection attacks: Data injection attacks involve injecting malicious or unauthorized data into the network or system monitored by the IDS. These attacks can disrupt the normal operation of the IDS by overwhelming it with excessive or irrelevant data, triggering false alarms, or causing system failures. Data injection attacks can exploit vulnerabilities in data handling mechanisms or target specific weaknesses in the IDS architecture.

Addressing adversarial attacks in IDS is a challenging task. Some strategies and techniques that can help to mitigate the impact of these attacks include:

- Adversarial training: Adversarial training involves training the IDS on both normal and adversarial examples to make it more robust against adversarial attacks. By exposing the IDS to various adversarial scenarios during training, it can learn to recognize and classify adversarial behavior more effectively.

- Defense mechanisms: Implementing defense mechanisms such as input sanitization, feature engineering, and anomaly detection ensembles can enhance the resilience of the IDS against adversarial attacks. These techniques focus on improving the robustness of the IDS to handle manipulated or malicious inputs.

- Monitoring and response: Continuous monitoring of the network and system activities can help to detect and respond to adversarial attacks in a timely manner. Real-time analysis, incident response, and adaptive countermeasures can aid in mitigating the impact of attacks and preventing further exploitation.

- Collaboration and information sharing: Sharing information and collaborating with other IDS systems, security researchers, and organizations can help to create a collective defense against adversarial attacks. Sharing knowledge about attack techniques, patterns, and countermeasures can lead to more effective defense strategies.

It is worth noting that adversarial attacks and defense mechanisms in IDS are evolving research areas, and new attack techniques and defense strategies are continuously being developed.

8. Emerging trends and future directions

1. Integration of deep learning techniques [34]: Deep learning techniques, such as deep neural networks and recurrent neural networks, have shown promising results in various domains. In anomaly detection for IDS, the integration of deep learning techniques can help to capture complex patterns and dependencies in network traffic data, improving the accuracy of detection [9].

2. Explainable AI for anomaly detection [35]: Explainability is a crucial aspect of anomaly detection in IDS. As complex machine learning models are being used, understanding and interpreting their decisions become essential. Future research focuses on developing explainable AI techniques that provide transparency and insights into the reasoning behind anomaly detections.

3. Real-time and streaming anomaly detection: Traditional batch processing approaches are not sufficient to handle the high-speed and large-scale nature of network traffic data. Future directions involve developing real-time and streaming anomaly detection methods that can process and analyze data on the fly, allowing for timely detection and response to anomalies.

4. Integration of multiple data sources: IDS can benefit from the integration of multiple data sources, such as network traffic data, system logs, and user behavior data. Incorporating diverse data sources and applying advanced fusion techniques can enhance the accuracy and robustness of anomaly detection.

9. Conclusion

In conclusion, this chapter has provided an overview of anomaly detection techniques in intrusion detection systems (IDS). We discussed the two main types of IDS, including signature-based detection and anomaly detection, and highlighted the advantages of using anomaly detection techniques over signature-based approaches. We explored various anomaly detection techniques, including statistical-based techniques, clustering-based techniques, density-based techniques, reconstruction-based techniques, and One-class support vector machines (SVM).

We also discussed the importance of data collection and preprocessing in IDS, emphasizing the relevance of different data sources and the need for effective preprocessing techniques to enhance anomaly detection accuracy. Furthermore, we covered the evaluation and performance metrics used to assess the effectiveness of anomaly detection techniques, including commonly used evaluation datasets and performance metrics such as accuracy, precision, recall, F1-score, ROC curve, and AUC.

We highlighted the challenges faced in anomaly detection for IDS, such as the curse of dimensionality, concept drift, and adversarial attacks. These challenges require ongoing research and development efforts to improve the accuracy and resilience of anomaly detection techniques. Additionally, we discussed emerging trends and future directions in the field, including the integration of deep learning techniques, the use of explainable AI, and the exploration of real-time and streaming anomaly detection methods.

In conclusion, anomaly detection techniques play a crucial role in IDS for enhancing network security by identifying potential threats and attacks. However, there are ongoing challenges and opportunities for further research and development. By addressing these challenges and embracing emerging trends, we can advance the field of anomaly detection in IDS and improve the detection and prevention of sophisticated and unknown attacks, ultimately enhancing the overall security of network systems.

Author details

Siamak Parhizkari
Islamic Azad University, Iran

*Address all correspondence to: parhizkari.siamak@live.com

References

[1] Kumar KN, Sukumaran S. A survey on network intrusion detection system techniques. International Journal of Advanced Technology and Engineering Exploration. 2018;**5**(47):385-393

[2] Modi C, Patel D, Borisaniya B, Patel H, Patel A, Rajarajan M. A survey of intrusion detection techniques in cloud. Journal of Network and Computer Applications. 2013;**36**(1):42-57

[3] Liu M, Xue Z, Xu X, Zhong C, Chen J. Host-based intrusion detection system with system calls: Review and future trends. ACM Computing Surveys (CSUR). 2018;**51**(5):1-36

[4] Khraisat A, Gondal I, Vamplew P, Kamruzzaman J. Survey of intrusion detection systems: Techniques, datasets and challenges. Cybersecurity. 2019; **2**(1):1-22

[5] Jyothsna V, Prasad R, Prasad KM. A review of anomaly based intrusion detection systems. International Journal of Computer Applications. 2011;**28**(7): 26-35

[6] Gangwar A, Sahu S. A survey on anomaly and signature based intrusion detection system (IDS). International Journal of Engineering Research and Applications. 2014;**4**(4):67-72

[7] Jmila H, Khedher MI. Adversarial machine learning for network intrusion detection: A comparative study. Computer Networks. 2022;**214**:109073

[8] Zamani M, Movahedi M. Machine Learning Techniques for Intrusion Detection. 2013. 11 p. Available from: arxiv.org [Revised in 2015]

[9] Kocher G, Kumar G. Machine learning and deep learning methods for intrusion detection systems: Recent developments and challenges. Soft Computing. 2021;**25**(15):9731-9763

[10] Yaacob AH, Tan IK, Chien SF, Tan HK. Arima based network anomaly detection. In: 2nd International Conference on Communication Software and Networks, 2010, Singapore. Singapore: IEEE; 2010. pp. 205-209

[11] Shirani P, Azgomi MA, Alrabaee S. A method for intrusion detection in web services based on time series. In: 28th IEEE Canadian Conference on Electrical and Computer Engineering, CCECE (CCECE). Halifax, Canada: IEEE; 2015. pp. 836-841

[12] Liu H, Lang B. Machine learning and deep learning methods for intrusion detection systems: A survey. Applied Sciences. 2019;**9**(20):4396

[13] Davis JJ, Clark AJ. Data preprocessing for anomaly based network intrusion detection: A review. Computers & Security. 2011;**30**(6–7): 353-375

[14] Alasadi SA, Bhaya WS. Review of data preprocessing techniques in data mining. Journal of Engineering and Applied Sciences. 2017;**12**(16):4102-4107

[15] Haq NF, Onik AR, Hridoy MAK, Rafni M, Shah FM, Farid DM. Application of machine learning approaches in intrusion detection system: A survey. IJARAI-International Journal of Advanced Research in Artificial Intelligence. 2015;**4**(3):9-18

[16] Salih AA, Abdulazeez AM. Evaluation of classification algorithms for intrusion detection system: A review. Journal of Soft Computing and Data Mining. 2021;**2**(1):31-40

[17] Aburomman AA, Reaz MBI. Survey of learning methods in intrusion detection systems. In: 2016 International Conference on Advances in Electrical, Electronic and Systems Engineering (ICAEES). Putrajaya, Malaysia: IEEE; 2016

[18] Bohara B, Bhuyan J, Wu F, Ding J. A survey on the use of data clustering for intrusion detection system in cybersecurity. International Journal of Network Security & Its Applications. 2020;**12**(1):1

[19] Wicaksana AK, Cahyani DE. Modification of a density-based spatial clustering algorithm for applications with noise for data reduction in intrusion detection systems. International Journal of Fuzzy Logic and Intelligent Systems. 2021;**21**(2):189-203

[20] Xu Y-X, Pang M, Feng J, Ting KM, Jiang Y, Zhou Z-H. Reconstruction-based anomaly detection with completely random forest. In: HAPPENING VIRTUALLY: SIAM International Conference on Data Mining (SDM21) April 29 - May 1, 2021, Virtual Conference. Philadelphia, PA, USA: SIAM; 2021

[21] Goodfellow I, Pouget-Abadie J, Mirza M, Xu B, Warde-Farley D, Ozair S, et al. Generative adversarial networks. Communications of the ACM. 2020;**63**(11):139-144

[22] Mahfouz AM, Abuhussein A, Venugopal D, Shiva SG. Network intrusion detection model using one-class support vector machine. In: Advances in Machine Learning and Computational Intelligence: Proceedings of ICMLCI 2019. Singapore: Springer Nature; 2021

[23] Panigrahi R, Borah S. A detailed analysis of CICIDS2017 dataset for designing intrusion detection systems. International journal of. Engineering & Technology. 2018;**7**(3.24):479-482

[24] Stiawan D, Idris MYB, Bamhdi AM, Budiarto R. CICIDS-2017 dataset feature analysis with information gain for anomaly detection. IEEE Access. 2020;**8**: 132911-132921

[25] Wang G, Hao J, Ma J, Huang L. A new approach to intrusion detection using artificial neural networks and fuzzy clustering. Expert Systems With Applications. 2010;**37**(9):6225-6232

[26] Parhizkari S, Menhaj MB, Sajedin A. A Cognitive Based Intrusion Detection System. 2020. 19 p. Available from: arxiv.org [Revised in 2022]

[27] Verleysen M, François D. The curse of dimensionality in data mining and time series prediction. In: Computational Intelligence and Bioinspired Systems: 8th International Work-Conference on Artificial Neural Networks, IWANN 2005, Vilanova i la Geltrú, Barcelona, Spain, June 8–10, 2005 Proceedings 8. Barcelona, Spain: Springer; 2005

[28] Aljanabi M, Ismail MA, Ali AH. Intrusion detection systems, issues, challenges, and needs. International Journal of Computational Intelligence Systems. 2021;**14**(1):560-571

[29] Brownlee J. Concept drift 2023. Available from: https://machinelearningmastery.com/gentle-introduction-concept-drift-machine-learning/

[30] Castillo D. what is concept drift 2023. Available from: https://www.seldon.io/machine-learning-concept-drift.

[31] Mbow M, Sakurai K, Koide H. Advances in adversarial attacks and defenses in intrusion detection system: A survey. In: Science of Cyber Security-

SciSec 2022 Workshops: AI-CryptoSec, TA-BC-NFT, and MathSci-Qsafe 2022, Matsue, Japan, August 10–12, 2022, Revised Selected Papers. Matsue, Japan: Springer; 2023

[32] Zizzo G, Hankin C, Maffeis S, Jones K. Adversarial attacks on time-series intrusion detection for industrial control systems. In: 2020 IEEE 19th International Conference on Trust, Security and Privacy in Computing and Communications (TrustCom) 29 Dec 2020 - 01 Jan 2021. Guangzhou, China: IEEE; 2020. ISBN: 978-0-7381-4380-4

[33] Alotaibi A, Rassam MA. Adversarial machine learning attacks against intrusion detection systems: A survey on strategies and defense. Future Internet. 2023;**15**(2):62

[34] Yehuda Y. New Trends in AI and Machine Learning for Anomaly Detection 2023. Available from: https://www.rad.com/blog/new-trends-ai-and-machine-learning-anomaly-detection

[35] Zehra S, Faseeha U, Syed HJ, Samad F, Ibrahim AO, Abulfaraj AW, et al. Machine learning-based anomaly detection in NFV: A comprehensive survey. Sensors. 2023;**23**(11):5340

Section 3

Anomaly Detection Models

Chapter 7

Verification of Generalizability in Software Log Anomaly Detection Models

Hironori Uchida, Keitaro Tominaga, Hideki Itai, Yujle Li and Yoshihisa Nakatoh

Abstract

With recent rapid technological advances, the automatic analysis of software logs has received particular attention. Currently, there is much research on the use of Deep Learning in the field of software log anomaly detection, and they have reported high accuracy of more than 0.9 in the f1-score. On the other hand, there are reports that it has not been used in the field of software development. We conducted a generalized evaluation against representative models for log anomaly detection to elucidate the cause of this problem. Five models were used in the subject: four representative models (two supervised and two unsupervised) and our proposed Neocortical Algorithm (supervised). We used the commonly used Blue Gene/L supercomputer log (BGL) dataset. The learning curves and cross-validation showed a tendency toward overfitting in all models. In addition, a survey of the frequency of log patterns confirmed the need for a more diverse dataset, as many of the patterns are a series of specific logs.

Keywords: anomaly detection, log analysis, log parsing, deep learning, software log, software development

1. Introduction

Automatic analysis of software logs has attracted significant attention due to the rapid development of technology in recent years. Currently, there are many studies with Deep Learning in the field of software log anomaly detection, reporting high accuracies surpassing 0.9 on the f1-score [1, 2]. On the other hand, it has been reported that Deep Learning for software log anomaly detection is not widely employed in the software development industry. The Loghub dataset, released by He et al., is presently frequently used in the field of software log anomaly analysis [3]. While Loghub contains logs from various systems, only one type of log is available for each system. Consequently, the accuracy of anomaly detection is assessed for only one pattern, and a comprehensive evaluation using multiple datasets is not performed. As a result, the effectiveness of the various anomaly detection models reported may be limited to specific datasets.

Therefore, to assess the generalizability of representative anomaly detection models across multiple dataset patterns, we initially conducted cross-validation using the Blue Gene/L supercomputer log (BGL) dataset from Loghub. For this purpose, we utilized the Deep-loglizer Toolkit developed by Chen et al. [4], which comprises four models, namely, CNN, LSTM (supervised learning), Transformer, and Auto Encoder (unsupervised learning). Additionally, we incorporated our proposed SPClassifier (supervised learning) to make use of a total of five models.

The second approach to evaluating generality involves using the validation datasets. In the study by Chen et al. that evaluates various models, the dataset is split into two partitions: the training dataset and the test dataset [1]. At each epoch, the models are evaluated on the test data, and the model with the highest accuracy in that epoch is considered the optimal model to calculate the accuracy for the test dataset. Given the potential for overfitting on the test dataset with this method, we split the dataset into three separate datasets for evaluation: the training dataset, the validation dataset, and the test dataset.

Furthermore, we examined the type and frequency of logs included in the dataset to assess whether the dataset is suitable for generic evaluation.

In summary, this experiment aims to clarify the following three points:

1. Evaluation of generality through cross-validation: Investigating the variation in accuracy due to differences in the types of logs included in the training and test datasets.

2. Evaluation of generality using the validation dataset: Assessing the generality using the validation dataset, which has not been included in previous benchmark studies.

3. Investigation of the log structure included in the dataset: Examining the similarity of the log structure in the commonly used BGL dataset to that used in software development.

2. Study design

This study uses the Toolkit (Deep-loglizer) provided by Chen et al. This Toolkit allows for flexibility in the model setup, including the ability to modify the loss function and determine whether or not to incorporate semantic information from the logs. We exclusively utilized sequential information in this experiment since our experimental setup lacked the necessary computational resources to handle semantic information. Notably, Chen et al. reported comparable accuracies with and without semantic information.

2.1 The common workflow

The common parts of the workflow for anomaly detection used in this experiment are shown in **Figure 1**. A standard anomaly detection model comprises four key steps: (1) Log Parsing, (2) Log Grouping, (3) Log Representation, and (4) Deep Learning Models [2]. Sections 2.1.1–2.1.4 provide an overview of each step and the specific techniques used in this study.

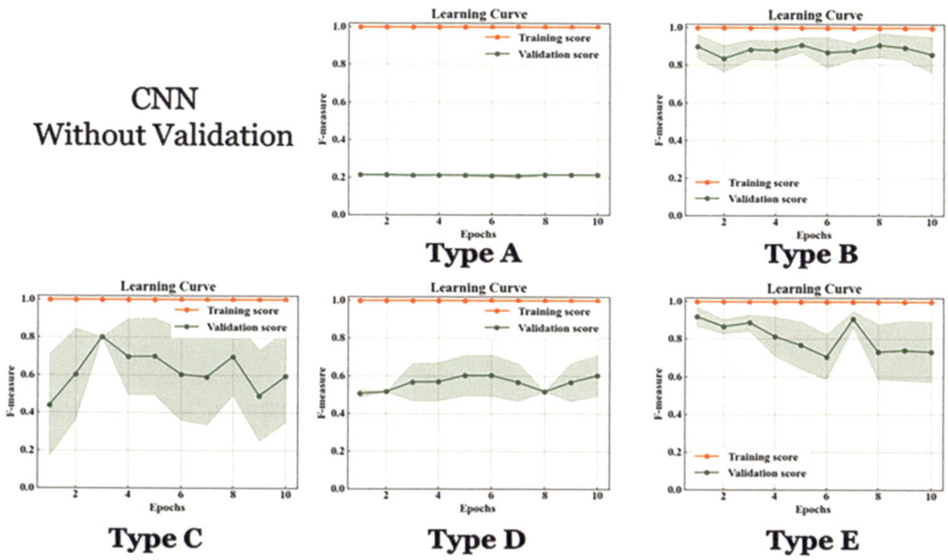

Figure 1.
Learning curve for CNN.

2.1.1 Log parsing

The log contains different fields such as timestamp, process ID, and severity. However, for the line-by-line anomaly label dataset, only the crucial log message portion is extracted and used through log analysis. Log analysis is used to automatically translate each log message into a specific event template (constant part) associated with the parameters (variable part). Various log analysis techniques are available, such as frequent pattern mining, clustering, heuristics, etc. For this experiment, log analysis was conducted using Drain (heuristic) [5], which has been used in benchmark studies. Drain uses fixed-depth analysis trees to speed up the analysis process and encodes rules specifically designed for the analysis. The parameters used are as follows: Similarity threshold = 0.5, Depth of all leaf nodes.

2.1.2 Log grouping

This step separates the logs into various groups. Typically, three types of windows are employed for log grouping, namely, Fixed Window, Sliding Window, and Session Window. Fixed Window is a grouping technique that splits the logs according to their frequency of occurrence, whereas Sliding Window segments the logs into window size and step size. Session Window, on the other hand, leverages log identifiers to group logs with the same execution path. For the current study, the Sliding Window of 10 and Sliding Step of 1 are utilized for log grouping.

2.1.3 Log representation

After grouping the logs, the logs are represented in the various formats required by the DL model. In general, they are converted into (1) sequential vectors, (2) quantitative vectors, and (3) semantic vectors. (1) Sequential vectors reflect the order of log events within a window. (2) Quantitative vectors use the frequency of occurrence of each log event within a log window. (3) Semantic vectors acquired from the language model represent the semantic feature of log events.

In this experiment, the (1) sequential vector is used.

2.1.4 Deep learning models

After the log representation step, the extracted features are fed into a deep-learning model for anomaly detection.

2.2 Evaluated models

In this experiment, we used a total of five models, consisting of two supervised learning models, namely Convolutional Neural Network (CNN) [6] and Long Short-Term Memory (LSTM) [7], and two unsupervised learning models, namely Auto Encoder (AE) [8] and Transformer [9], in addition to our proposed SPClassifier [10]. An overview of each model and the Toolkit parameters used in this study is provided in Section 2.2.1–2.2.5.

2.2.1 CNN (supervised model)

The input logs undergo the following preprocessing steps: first, they are converted to IDs, then to vectors using logkey2vec, and finally fed into the CNN model. This approach resulted in an impressive F-measure of 0.98 on the HDFS dataset. The specific Toolkit parameters used for this experiment are as follows: "python cnn_demo.py –label_type anomaly –feature_type sequential – 10."

2.2.2 LSTM (supervised model)

This approach utilizes fixed-dimensional semantic vectors to represent log events and employs an attention-based Bidirectional Long Short-Term Memory (Bi-LSTM) classification model for anomaly detection. The Toolkit parameters used for this experiment are as follows: "python lstm_demo.py –label_type anomaly –feature_type sequential –use_attention –topk 50 –epochs 10".

2.2.3 Auto encoder (unsupervised model)

The model consists of two Auto Encoders and one Isolation Forest; the Auto Encoder is used for feature extraction and anomaly detection, and the Isolation Forest is used for positive sample prediction. The parameters used in the Toolkit for this experiment are as follows: "python ae_demo.py –feature_type sequential –anomaly_ratio 0.8 –epochs 10."

2.2.4 Transformer (unsupervised model)

Existing approaches exhibit limitations in their ability to generalize to new, unseen log samples. To address this issue, Logsy was proposed as a novel method for anomaly detection, utilizing a self-attention encoder network for hyperspherical classification. Logsy formulates the log anomaly detection problem by discriminating between regular training data from the target system and samples from auxiliary log datasets that are easily accessible from other systems.

The parameters used in the Toolkit for this experiment are as follows: "python transformer_demo.py –label_type next_log –feature_type sequential –topk 50 –epochs 10 –use_attention."

2.2.5 SPClassifier (supervised model)

SPClassifier is a model with sparse features and internal representations suitable for training in CPU environments [10]. The proposed method consists of one spatial pooling layer [11, 12] and one Feedforward Neural Network for classification, which identifies anomalous patterns from log data transformed into 2D features. The feature transformation process involves converting the input log sequence into a sparse distributed representation (SDR), which is a binary sequence of fixed dimensions. In the SDR, a specific percentage of the bits are set to 1, while the remaining bits are set to 0. These transformed features serve as input to the spatial pooling stage. Spatial pooling (SP) incorporates local suppression between adjacent mini-columns and implements a k-wins-take-all computation. At any given time, only a small fraction of the mini-columns with the most active inputs are active. Feed-forward connections to active cells are adjusted at each time step based on Hebb's learning rule. Additionally, a homeostatic excitation control mechanism, referred to as "boosting," operates on a slower time scale. Boosting enhances the relative excitability of underactive mini-columns, encouraging their activation and participation in the input representation. Subsequently, the transformed SDRs obtained through spatial pooling are fed into a classifier responsible for detecting anomalies. The employed classifier utilizes a single-layer neural network. It takes as input a flat binary SDRs array representing the output of the spatial pooling layer and predicts the abnormal or normal label. A softmax function is employed as the activation function for the output of the network. The network weights are updated during training using a formula based on the provided label information.

$$w_{ij} \leftarrow w_{ij} + \alpha \times \sum_{i=0}^{c-1} \left\{ \frac{1}{c} - softmax\left(\sum_{i=0}^{c-1} w_{ij} z_j \right) \right\} \tag{1}$$

where w_{ij} is the weight between the j-th value of the flattened input z_j and the i-th output node of the neural network. c is the number of categories, c = 2 since normal and abnormal categories are used for anomaly detection. α is a coefficient that controls the speed of learning.

The parameters are shown in **Table 1**.

Parameter	Value	Description
Window size	10	Size of sliding window
Stride	1	Step size of sliding window
Input SDR Length	500	Number of dimensions of SDR transformation for a single template index.
Input SDR Sparsity	0.15	Proportion of binary values equal to 1 across all dimensions post SDR conversion.

Parameter	Value	Description
Input Dimensions	(500, 10)	Shape of the coded image generated by stacking SDRs.
Column Dimension	(830, 15)	Shape of the columns in spatial pooling layer.
Potential Radius	7	Value that determines the input range over which one column has a potential connection.
Potential Pct	0.1	Percentage of inputs with potential connections in the hypercube.
Global Inhibition	True	Whether to consider all columns as neighbors when determining the active state of a column.
Local Area Density	0.1	Percentage of columns that can be active between neighbors.
Stimulus Threshold	6	Minimum number of synaptic connections required for a column to be active.
SynPermActiveInc	0.14	Amount of increase in permanence value of active synapses at each learning step.
SynPermInactiveDec	0.02	Amount by which the permanence value of inactive synapses decreases with each learning step.
SynPermConnected	0.5	Minimum permanence value at which a synapse is considered connected.
DutyCyclePeriod	1402	Length of time step considered when updating the boost factor based on how often each column is active.
MinPctOverDutyCycles	0.2	Lower limit on how often a column has active input above the stimulus threshold.
Boost Strength	7	Parameters that control the strength of the boost factor's adaptive effect.
WrapAround	False	Whether the first and last dimensions of the input are considered adjacent in the mapping between input and column.

Table 1.
Parameter list for SPClassifier.

2.3 Dataset

The BGL dataset collected from the supercomputer system Blue Gene/L was used as the evaluation dataset for this experiment. This dataset contains log data labeled with anomaly logs. It was sourced from Loghub [3], a renowned repository offering a vast collection of log datasets that can be employed for AI-based log analysis.

2.3.1 Data selection strategies

In this study, a set of time series was used, where 80% of the available logs were allocated for training and the remaining 20% for testing. The testing dataset was further split into two halves, with the initial half used for validation purposes and the second half used for testing. Such a partitioning strategy emulates a practical software development scenario, where past logs are utilized for training and future logs are used for testing purposes.

2.3.2 Different data grouping

There are two major data grouping techniques: window grouping and session grouping. In this experiment, we used the window grouping approach, utilizing a window size of 10 with a sliding value of 1. It is worth noting that the results obtained using session windows outperform those obtained using fixed windows on the BGL dataset, as previously reported by Le et al.. This result may be attributed to the fact that the larger size of the input data augments the amount of information that can be acquired, which facilitates the detection of anomalies. In this study, we selected to use fixed grouping with a window size of 10, as we believe that anomaly detection at a more detailed level is important, particularly in a developmental setting.

2.4 Evaluation metrics

2.4.1 Evaluation method

In the accuracy comparison, each model after training is used to verify the accuracy of anomaly detection for test data. The classification performance of each model is evaluated by the F-measure value; F-measure is an evaluation index that indicates the balance between detection accuracy and the number of anomalies detected. Here, the F-measure is calculated as follows.

$$Precision = \frac{TP}{TP + FP} \tag{2}$$

$$Recall = \frac{TP}{TP + FN} \tag{3}$$

$$F - measure = \frac{2 \bullet Precision \bullet Recall}{Precision + Recall} \tag{4}$$

where:
TP represents abnormal instances correctly classified by the model.
TN represents normal instances correctly classified by the model.
FP represents normal instances misclassified by the model.
FN represents abnormal instances misclassified by the model.

2.4.2 Learning curve

Each training set consisted of one epoch, and during training the model was evaluated for accuracy using both training and validation data. The training range consists of epochs 1 through 10. In experiments without a validation set, the test data were evaluated as the validation set.

2.5 Objectives of the experiment

Objective 1: Generalizability evaluation with cross-validation.

To evaluate the generalizability of each model, we employed five-fold cross-validation in this experiment. Four-fifths of the dataset was used as the training dataset, while the remaining one-fifth was designated as the test dataset. The test dataset was subsequently split in half, with the first half serving as the validation dataset. By varying the split points and comparing the accuracy of anomaly detection, we will explore the potential impact of training dataset bias and test dataset bias on the performance of the system.

Objective 2: Generalizability evaluation with the validation dataset.

This experiment aims to evaluate the accuracy with and without the validation dataset, as previous studies only split the dataset into training and test data. The test data are used to determine the optimal model among multiple Epoch training runs; however, there is a potential risk of over-training on the test data. Hence, this study incorporates the validation dataset to assess whether over-training occurs. The dataset segmentation method was the same as Obj1, with train 80%, validation 10%, and test 10%. Our previous studies have shown that the accuracy is extremely high when the train dataset ratio is increased to 90% [13]. We attribute this result to the extreme reduction in the number of unknown anomaly logs not present in the training dataset, from 268 to 3, resulting in fewer opportunities to test the unknown anomaly logs. In the actual field of development, the source code is updated daily and the logs are updated accordingly. Therefore, we use 80% of the training data, a condition that includes a large number of unknown anomaly logs, to measure the generality of the test.

Objective 3: Examination of Log Structure within the dataset.

In this section, we explore the dataset following the grouping of raw logs into fixed groups with a window size of 10. It is essential to examine the types of logs contained in the dataset to evaluate its versatility. For example, an application development site may receive more than 10,000 different types of logs. Therefore, if the experimental dataset comprises only 100 log types, it is inadequate in representing an actual development site, and the accuracy of anomaly detection will be questionable. Thus, this experiment aims to determine the number of log types and their frequency within the dataset.

3. Experimental results

We conducted five trials to eliminate errors in each experiment and evaluated the results using their average.

3.1 Obj1: Generalizability evaluation with cross-validation

Table 2 presents the results obtained through cross-verification applied to the five models. The table reveals the following four observations:

Types of cross verification	F-measure without/with	Recall	Precision
	CNN		
Type A	0.217/0.216	0.124/0.122	0.886/0.945
Type B	0.961/0.925	0.992/0.971	0.932/0.885
Type C	0.837/0.698	0.961/0.962	0.750/0.581
Type D	0.718/0.514	0.699/0.151	0.777/0.040
Type E	0.971/0.755	0.996/0.996	0.949/0.636
Ave Without/With	0.741/0.622	0.755/0.683	0.859/0.782
	LSTM		
Type A	0.211/0.207	0.118/0.116	0.981/0.975
Type B	0.928/0.908	0.986/0.949	0.883/0.879
Type C	0.540/0.392	0.959/0.963	0.422/0.272
Type D	0.771/0.516	0.784/0.365	0.756/0.877

Types of cross verification	F-measure without/with	Recall	Precision
Type E	0.942/0. 909	0.994/0. 997	0.895/0.837
Ave Without/With	0.678/0.586	0.768/0.678	0.788/0.768
	SPClassifier		
Type A	0.176/0. 514	0.279/0. 427	0.595/0. 696
Type B	0.759/0.689	0.843/0.932	0.695/0.549
Type C	0.568/0.546	0.624/0.777	0.609/0.456
Type D	0.569/0.181	0.426/0.115	0.893/0.675
Type E	0.871/0.840	0.948/0.966	0.809/0.744
Ave Without/With	0.589/0.554	0.624/0.643	0.720/0.634
	Auto encoder		
Type A	0.160/0.118	0.954/0.588	0.087/0.080
Type B	0.226/0.214	0.982/0.975	0.128/0.120
Type C	0.019/0. 007	0.850/0. 669	0.010/0. 004
Type D	0.400/0. 315	0.666/0. 848	0.306/0. 218
Type E	0.362/0. 152	0.962/0. 994	0.232/0. 083
Ave Without/With	0.233/0.161	0.883/0.815	0.152/0.101
	Transformer		
Type A	0.159/0.159	0.963/0.961	0.087/0.087
Type B	0.457/0.429	0.709/0.712	0.338/0.307
Type C	0.257/0.243	0.521/0.451	0.172/0.166
Type D	0.587/0.508	0.519/0.484	0.675/0.562
Type E	0.865/0.810	0.911/0.911	0.824/0.744
Ave Without/With	0.465/0.430	0.725/0.704	0.419/0.373

Table 2.
Generalizability evaluation results by cross-validation.

1. For all models, the accuracy is notably low when using the Type A dataset.

2. For all models, the accuracy is good when using Type B and Type E datasets.

3. When using the Type D dataset, the accuracy diminishes in supervised learning, whereas unsupervised learning demonstrates higher accuracy compared to the other four types of datasets.

4. When Type A dataset is used, unsupervised learning results in high precision and low recall, while unsupervised learning results in high precision and low recall.

We attribute observation (4) to the contrasting characteristics of the two learning methods. Supervised learning involves learning from the training data one-to-one, making it proficient in correctly identifying the anomalous data on which it is trained.

Types of cross verification	Normal types Test/validation	Anomaly types Test/validation	Total anomaly logs Test/validation
Type A	681/263	9/8	24,284/ 6881
Type B	64 /27	0 /1	0 /24
Type C	26 / 9	1 /0	24 / 0
Type D	8 /15	0 /0	0 /0
Type E	18 /98	0 /0	0 /0

Table 3.
Composition of unknown logs included in each dataset used for cross-validation.

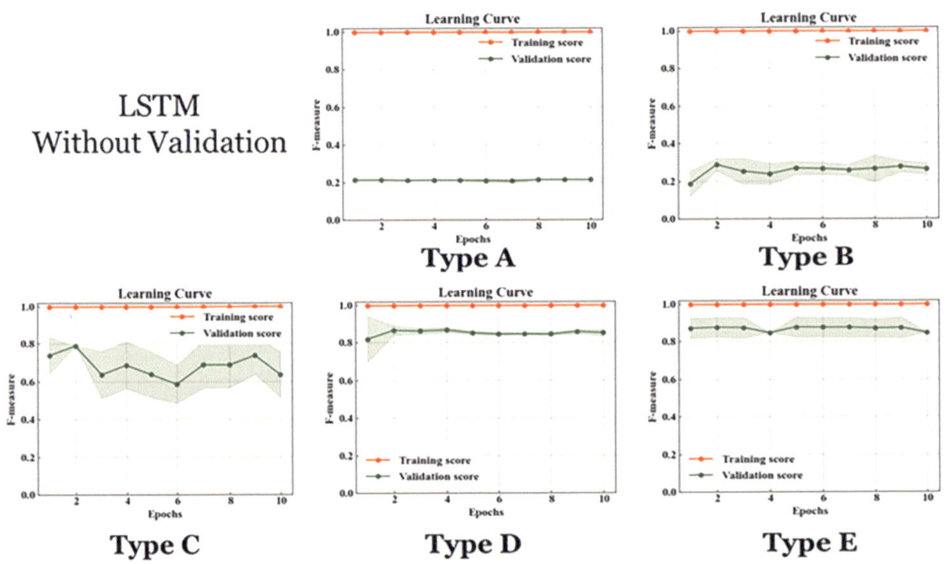

Figure 2.
Learning curve for LSTM.

However, it struggles to detect anomalous patterns outside its learning range, leading to a significant drop in Recall. On the other hand, unsupervised learning often detects patterns that deviate from the training data as abnormal, resulting in higher recall. However, it also tends to identify non-anomalous data as abnormal, leading to a substantially lower precision.

Figures 1 and **2** show the learning curves of the CNN and LSTM models, showcasing their high accuracy. The data points on the graph represent the mean values obtained from five experiments, with the upper and lower widths representing the standard deviations. **Figure 1** demonstrates a substantial variation for Type C, Type D, and Type E.

These findings indicate that the cross-validation results exhibit significant variations in accuracy across different types, suggesting limited generality and versatility.

As an additional investigation, we examined the number of unknown normal and abnormal logs in each split type. The results are presented in **Table 3**. The table reveals that the Type A dataset, which yielded the poorest results, had the highest number and diversity of abnormal logs. Conversely, the datasets used for testing

Type B, Type C, and Type D, which exhibited relatively good results, did not contain any unknown anomaly logs. This suggests that the supervised learning method achieved high accuracy primarily because it was tested with learned anomaly logs. These findings highlight the susceptibility of the system to unknown anomaly logs and indicate that it may not be universally applicable for certain purposes.

3.2 Obj2: Generalizability evaluation with the validation dataset

The comparison between the cases with and without the validation dataset, as shown in **Table 2**, reveals the following characteristics:

1. The CNN, LSTM, and AE models show a decrease in accuracy when the validation dataset is utilized.

2. The Transformer model maintains its accuracy even when the validation dataset is used.

3. The SPClassifier model maintains a basic level of accuracy, but its performance varies depending on the dataset type. Accuracy increases when Type A is used and decreases when Type D is used.

Especially, the accuracy of the CNN and LSTM models decreased by 0.2 when the validation dataset was employed, indicating a potential issue of over-training during the validation phase. Furthermore, the precision values for both CNN and LSTM significantly dropped when validation was used, resulting in a higher number of instances where anomalous data were incorrectly identified as normal. This suggests that over-training may occur during the validation stage, potentially impeding the models' ability to accurately detect anomalies in the test dataset.

The reason behind the observed characteristic (3) can be attributed to the learning method employed by the SPClassifier. In this method, the Spatial Pooling layer generates similar firing patterns for similar inputs and distinct firing patterns for different inputs. Consequently, we posit that utilizing a more complex dataset for validation would lead to a more refined Classifier threshold, ultimately enhancing the accuracy of the model.

3.3 Obj3: examination of log structure within the dataset

The experimental dataset was constructed from the BGL dataset using a sliding grouping method (Window size = 10, Sliding = 1). Sequence Patterns were formed based on templates delimited by Window, and the survey focuses on identifying the number of distinct Sequence Patterns present. **Table 3** showcases a total of 19 Sequence Patterns, which account for 80.3% of the dataset. Considering that there are 131,803 Sequence Patterns in total, the fact that the top 19 patterns represent such a significant portion indicates a bias toward specific Sequence Patterns. It is noteworthy that these 19 patterns are composed of only three templates, suggesting successive repetitions of the same log.

While such Sequence Patterns are commonly observed in OS system logs and network system logs, they are less prevalent in application development, which constitutes a significant portion of software development. In large-scale application development scenarios, where there are tens of thousands of diverse logs, the simultaneous operation of multiple systems leads to the appearance of complex Sequence Patterns in the output logs. Consequently, a model that exhibits high accuracy on a BGL dataset like this one may not achieve the same level of accuracy when applied to application development due to the inherent differences in the log patterns.

4. Conclusion

In this experiment, our focus was to investigate the limited adoption of deep neural network (DNN)-based anomaly detection methods in the development field. Existing anomaly detection models tend to classify anomaly logs as normal when window grouping is applied. Additionally, when incorporating validation data, the models tend to overfit and exhibit stable learning curves from the initial epoch.

Furthermore, we delved into the structure of the BGL dataset employed in this experiment and observed that certain logs appeared consecutively, with specific Sequence Patterns accounting for a substantial portion of the dataset. In addition, we performed a more in-depth examination of the structure of the BGL dataset used in this experiment. Our findings revealed a recurring occurrence of specific logs within the BGL dataset, along with the presence of certain Sequence Patterns that encompass a significant fraction of the logs. It is crucial to note that in application development, which constitutes a substantial aspect of software development, logs exhibit a greater level of complexity and encompass a wide range of diverse Sequence Patterns. Consequently, the existing representative model faces challenges when applied to the realm of application development.

While the anomaly detection field often focuses on logs with repeated occurrences, such as Super Computer logs or network systems, we aim to target anomaly detection in logs associated with large-scale software development, including applications. Therefore, our plans involve creating diverse datasets that reflect the characteristics of the field under development and exploring the feasibility of employing multiple anomaly detection systems in this context.

Acknowledgements

This work is supported by a grant from Panasonic System Design.

Author details

Hironori Uchida[1*], Keitaro Tominaga[2], Hideki Itai[2], Yujle Li[1] and Yoshihisa Nakatoh[1]

1 Kyushu Institute of Technology, Fukuoka, Japan

2 Panasonic System Design Co., Ltd., Kanagawa, Japan

*Address all correspondence to: uchida.hironori182@mail.kyutech.jp

References

[1] Chen Z, Liu J, Gu W, Su Y, Lyu M. Experience Report: Deep Learning-based System Log Analysis for Anomaly Detection. United States. Arxiv: https://arxiv.org/pdf/2107.05908.pdf [Accessed: 30 April 2023]

[2] Le V, Zhang H. Log-based anomaly detection with deep learning: How far are we? In: ICSE '22: Proceedings of the 44th International Conference on Software Engineering. Software Engineering. United States: Association for Computing Machinery; 2022. pp. 1356-1367

[3] He S, Zhu J, He P, Lyu MR. Loghub: A Large Collection of System Log Datasets towards Automated Log Analytics. United States. 2020. Arxiv Website: https://arxiv.org/pdf/2008.06448.pdf [Accessed: 30 April 2023]

[4] Deep-loglizer. Available from: https://github.com/logpai/deep-loglizer [Accessed: 30 April 2023]

[5] He P, Zhu J, Lyu MR. Drain: An Online Log Parsing Approach with Fixed Depth Tree. 2017 IEEE International Conference on Web Services (ICWS)

[6] Lu S, Wei X, Li Y, Wang L. Detecting Anomaly in Big Data System Logs Using Convolutional Neural Network. 2018 IEEE 16th Intl Conf on D; 2018

[7] Zhang X, Xu Y, Zhang H, Dang Y, Xie C, Yang X, et al. Robust log-based anomaly detection on unstable log data. In: ESEC/FSE 2019: Proceedings of the 2019 27th ACM Joint Meeting on European Software Engineering Conference and Symposium on the Foundations of Software Engineering. Software Engineering. United States: Association for Computing Machinery; 2019. pp. 807-817

[8] Farzad A, Gulliver TA. Unsupervised log message anomaly detection. ICT Express; 2020;**6**:229-237

[9] Nedelkoski S, Bogatinovski J, Acker A, Cardoso J, Kao O. "Self-Attentive Classification-Based Anomaly Detection in Unstructured Logs" 2020 IEEE International Conference on Data Mining (ICDM). Italy: IEEE; 2020

[10] Hirakawa R, Uchida H, Nakano A, Tominaga K, Nakatoh Y. Large scale log anomaly detection via spatial pooling. Cognitive Robotics. Oct 2021;**1**:188-196. DOI: 10.1016/j.cogr.2021.10.001

[11] Cui Y, Ahmad S, Hawkins J. "The HTM spatial pooler—A neocortical algorithm for online sparse distributed coding." Frontiers in Computational Neuroscience. Nov 2017;**11**. DOI: 10.3389/fncom.2017.00111

[12] Li L, Zou T, Cai T, Niu T, Zhu Y. A fast spatial Pool learning algorithm of hierarchical temporal memory based on Minicolumn's self-nomination. Computational Intelligence and Neuroscience. 2021;**2021**. DOI: 10.1155/2021/6680833

[13] Uchida H, Tominaga K, Itai H, Li Y, Nakatoh Y. Investigation of Weaknesses in Typically Anomaly Detection Methods for Software Development., IHIET2023. (in press)